Exact Values in Trigonometry

Five New Techniques

Bhava Nath Dahal

Copyright © 2016 Bhava Nath Dahal
All rights reserved.

ISBN -10: 1536995002
ISBN-13: 978-1536995008

My words

During the study of classical astronomy, I found, there are different style of interpretation of a circle. This leads me; there would be something undiscovered special characteristics of circle to us. This inspired me to search the trigonometric ratios.

I am not a mathematician. Therefore, this is not a professional assignment, rather just an amateur. During development of the new method for trigonometric relations, I faced un-numbered problems of algebra, geometry and arithmetic. I tried to resolve using various literature, webpages and contacted possible persons and online-forum. Many of the contacted persons or forum could not assist this hobby task. Few of them assisted a bit which were valuable to me. I thankfully acknowledge the assistance and cooperation from them.

There are five different methods- one is ratio-based and four are chord-based. Four methods gives exact trigonometric values. These are new concept of an amateur; therefore have much room for professionally improvement. Nonetheless, it opens new interpretation for classical method of trigonometry. A good mathematician or mathematical institute may assist to improve its quality. The language and its flow in the book has room to improve also.

Readers are the judge for any art. Your comments, suggestions and academic discussions are most welcome for further improvement.

Bhava Nath Dahal
Kathmandu, Nepal
2016 August 15

To

Dāsu Bhātta, Shivāram Pādhyā, Bashudev Pādhyā, Benumadhāv

Pādhyā, Tulāram Pādhyā, Shankār Pādhyā,

Kāshināth, Hirāmani, Govindā, Gobardhān,

Umāpati, Srināth, Govindānath,

Bhimnāth – Kāusalyā Dahal

Table of Content

CHAPTER 1   TRIGONOMETRIC RATIO

1.1	Trigonometric Ratios	1
1.2	Historical development	3
1.3	Geometric Method	6
1.4	Chord and Supplementary Chord	9
	1.4.1 Half angle Chords	11
	1.4.2 Double angle Chords	13
	1.4.3 Triple angle Chords	14
	1.4.4 Addition and subtraction of Chords	14
	1.4.5 Calculated Chords	15
1.5	Sine Square	17
1.6	Sine Scale	19

CHAPTER 2   ARC-LINE METHOD

2.1	Arc-Line concept	21
	2.1.1 Inputs	22
	2.1.2 Accuracy	22
	2.1.3 Mathematical process	24
2.2	Trigonometric Ratios of any angle	27
2.3	Angle-based Formula	28

CHAPTER 3   SERIES-METHOD

3.1	Exact Sine from Series Method	29
	3.1.1 Angle and Complementary-angle Series	30
	3.1.2 Sine of $A/2^n$-Series	40

CHAPTER 4		**ANGLE-REWRITTEN METHOD**	
4.1	Exact Sine of an Angle		47
	4.1.1	Half-Series	49
	4.1.2	Varieties of Techniques for Angle-Rewritten	52
4.2	General procedures - exact value		57
CHAPTER 5		**PRECISE-REWRITTEN METHOD**	
5.1	Precise-Rewritten Concept		65
	5.1.1	Inputs	66
	5.1.2	Accuracy	67
	5.1.3	Mathematical process	69
5.2	Sine of any angle		74
CHAPTER 6		**VATSA METHOD**	
6.1	Ratio-based Sine Table		79
	6.1.1	For Sexagesimal System (Degrees)	79
	6.1.2	For Metric System (Grades)	84
6.2	Accuracy		88
6.3	Sine of an Individual Angle		89
CHAPTER 7		**VASTA- NATURAL NUMBER SERIES**	
7.1	New Method for Series-Sum		91
	7.1.1	First Series (Sum of Arithmetic Series)	91
	7.1.2	N_{th} Series	93
7.2	Sum of Series of Sequence		95

Chapter 1
TRIGONOMETRIC RATIO

1.1 Trigonometric Ratios

The length of perpendicular from the vertex of an angle of a triangle to the opposite base is Sine of opposite angle of that perpendicular. In the modern mathematics, this is deemed as ratio of perpendicular to hypotenuse of a right angle triangle. Based on modern perception, there are six trigonometric ratios- Sine, Cosine, Tangent, Cotangent, Secant and Cosecant for an angle. For written text, they are written as Sin A, Cos A, Tan A, Cot A, Sec A, CSC A for an angle named as A. Many people use variations mainly on CSC A as Cosec A or use 2-digit notations for all trigonometric relations.

EXACT VALUES IN TRIGONOMETRY

Apart from above six main ratios, there are less-common-use 10-ratios as Versine (Utkrāmajyā), Vercosine, Coversine, Covercosine, Exsecant, Excosecant, Haversine, Havercosine, Hacoversine and Hacovercosine. Chord (a) of an angle, its supplementary chord (b) or their product (ab) make all of the trigonometric ratios as:

Main relations:

$\sin A = ab/2$ $\qquad$ $\csc A = 2/ab$

$\cos A = 1 - a^2/2$ $\qquad$ $\sec A = \dfrac{2}{2-a^2}$

$\tan A = \dfrac{ab}{2-a^2}$ $\qquad$ $\cot A = \dfrac{2-a^2}{ab}$

Verse and Coverse relations:

$\text{Versin } A = a^2/2$ $\qquad$ $\text{Vercos } A = b^2/2$

$\text{Coversin } A = 1 - ab/2$ $\qquad$ $\text{Covercos } A = 1 + ab/2$

$\text{Exsec } A = \dfrac{a^2}{2-a^2}$ $\qquad$ $\text{Excsc } A = (2-ab)/ab$

Haverse and Hacoverse relations:

$\text{Haversine } (a^2/4)$ $\qquad$ $\text{Havercosine } (b^2/4)$

$\text{Hacoversine } (2-ab)/4$ $\qquad$ $\text{Hacovercosine } (2-ab)/4$

If we know the length of chord, half of that chord (a/2) will be the Sine of half of angle representing that chord (not included above). This is the magical development of determination of Sine of an angle since last 2000 years.

Many mathematicians' worked for resolve of trigonometric ratios in 2000-2500 years or before. They developed variety of methods and interpolation techniques. Arabs, Greek and Hindu Mathematicians made major contributions. They developed and preserved the methods to us. In the latest calculators and computers are available for calculations.

TRIGONOMETRIC RATIOS

The determination of trigonometric ratio, determination of Sine of an angle is important. Other ratios can be, somewhat, easily calculated based on value of Sine. This book is elaborating only the Sine of an angle, mainly in the length basis of a unit circle (having radius of 1).

1.2 Historical development

Hindu's Sine Table

Hindus developed their Sine table probably around 5th – 8th Century BC during *Sulba Sutram* period. Their main formula is 96th part of a circle seems (straight) as stick.

One interesting thing with Hindu Mathematicians, they did not know the relation of angles. In their geometry, angles were handled through value of Sine to that angle *(ex*act name is Jiva[1], used as *Jya or Jyardha*) in terms of length. Hence, Sine of an angle in length (*Jya*) was vital for their mathematics. Secondly, they have not unique number for π. Hindus took different radius for a circle. Most mathematicians since 5th Century took radius of 3438 for a circle having circumference 21600[2].

---

[1] The word 'Sine' is based on Jiva (still commonly usable in Nepal, India, Bangladesh and other part of South Asia) which is transferred to Europe through Arabs: see https://en.wikipedia.org/wiki/Sine. The exact word *Jiva* (meaning chord) transferred to Arabians spelt as *jayb* (meaning bosom). The meaning had transferred to Latin as *Sinus*, now Sine.

[2] This circumference was based on units of a day (1 day= 60 ghatikas= 3600 palas = 21600 vipalas) which was equivalent to one circle from the geocentric Sun (360 anshs= 21600 kalas).

EXACT VALUES IN TRIGONOMETRY

Guru Aryabhatiya, Guru Bramhagupta, Guru Bhaskara used the Jya-system for determination of Sine as 24-Jya, 10-*laghujya etc*. Surya Sidhhanta describes the same formula for 24 *Jya*. It started from 3.75° and increased by same upto 90°. Their presumption of 3.45° is a straight line. For a circle having circumference of 21600 (units) with 3438 radius, first *Jya* (length of perpendicular for Sin 3.45°) shall[3] be 225. From the first *jya*, second *jya* (length of perpendicular for Sin 7.5°) is 225+ (225-225/225) = 449. Similarly, third Jya (length of perpendicular for Sin 11.25°) is 449+224 − (449/225) =671 and so on.

Greek/Roman Sine Table

Guru Euclid (Father of Geometry - 3rd Century BC) developed Sine Table based on his formula of product of two diagonal is equal to the sum of product of opposite sides of quadrilateral in the circumference. He was the first major contributory for circle and triangle. His chords were in each ½ degree of a circle. Almost trigonometric function is based on his works until now.

Guru Pythagoras (580-500 BC) contributed the circle and triangles including extremely usable formula is relation of sides with hypotenuse.

Guru Euler (Leonhard Euler- 1707-1783) worked mainly in triangles and less in Sine-table. Euler-line is in his respect.

---

[3] This is 21600/4= 5400 in a quarter of a circle. There are 24 *Jya* in a quarter of a circle, so first *Jya* is 225 (=5400/24). In other way, 21600/96=225 seems as straight as per *Sulba-Sutram*. Suryasidhhanta takes 30°×60'/8=225 for this purpose. In all cases, the principle adopted is same as derived from *Sulba-Sutram*.

TRIGONOMETRIC RATIOS

Guru Taylor (Brook Taylor- 1685-1731) developed his Sine-formula based on radian of measurement of angle in 1715. For any angle of x radians[4],

$$Sin\ x = \frac{x}{1!} - \frac{x^3}{3!} + \frac{x^5}{5!} - \frac{x^7}{7!} + \cdots$$

This formula can be rewritten as:

$$Sin\ x = \frac{x}{1} - \frac{x^3}{6} + \frac{x^5}{120} - \frac{x^7}{5040} + \cdots$$

User of degree of measurement need to convert degrees into radians dividing by 180 and multiplying by π.

Answer from this method shall be accurate almost upto 7 digits. This is not a separate table of Sine, rather, part of Taylor's series give values for Sine.

Numerous respectable mathematicians contributed for trigonometry and geometry, so we can achieve modern level.

---

[4] Exactly same formula had developed by Madhava (1350-1425). Madhava's notation had based on Hindu mathematics in arc minute, arc second system instead of modern radian.

EXACT VALUES IN TRIGONOMETRY

1.3 Geometric Method

Some of the angle gives its value of Sine through geometric method. All these methods were developed and used by respected historical mathematicians for us.

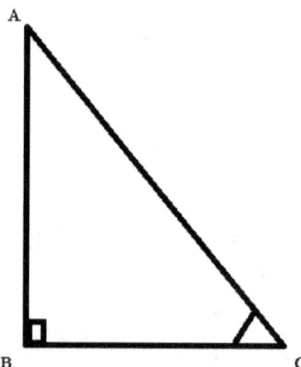

In the Triangle ABC, Sin C = AB/AC and Cos C = BC/AC.

Sin 90° (Sin $\frac{\pi}{2}$)

In 90°, opposite side and hypotenuse deemed to be same. So,

Sin 90° = 1

> *For 90°, chord is √2 and supplementary chord is also √2.*

Alternatively, we can take half of double angle chord for 90° in a circle. Half-chord is sine of half-angle has discovered since before 2000 years. Double of 90° is 180°, which has chord having length equal to diameter (2 in case of a unit circle). Halving the angle and chord both, Sin 90°= 2/2=1.

Sin 45o (Sin $\frac{\pi}{4}$)

In a right-angled triangle, if one of the angle is 45°, other become 45°. So, their base and perpendicular (adjacent and

TRIGONOMETRIC RATIOS

opposite sides) shall be same. Based on Guru Pythagoras, $p^2+b^2=h^2$, and $p=b$,

$p^2+p^2=h^2$,

$h^2 = 2p^2$

So, $h=\sqrt{2}p$.

Hence, $\sin 45° = p/h = p/\sqrt{2}p = \dfrac{1}{\sqrt{2}}$

Alternatively, we can take half of double angle chord for 45° in a circle. Double of 45° is 90°, which has chord of √2 (in a unit circle based on Guru Pythagoras). Halving the angle and chord both, $\sin 45° = \sqrt{2}/2$.

Sin 60o (Sin $\frac{\pi}{3}$) and Sin 30o (Sin $\frac{\pi}{6}$)

In a right-angled triangle, if one of the angle is 60º, other become 30º. If we extended perpendicular line to mirror triangle at the center, it becomes isosceles triangle having same length in each side.

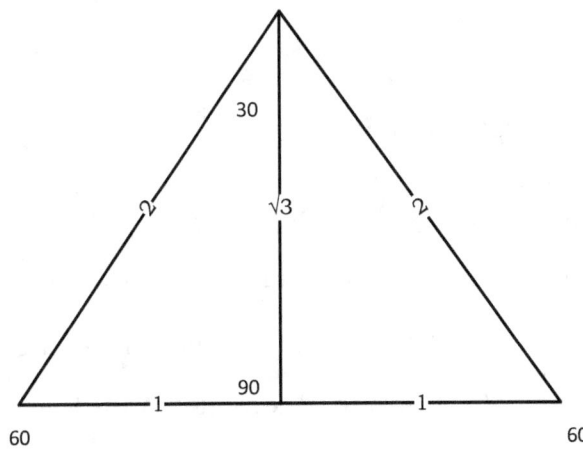

Therefore, if the base in right-side triangle is 1, hypotenuse will be 2 and perpendicular becomes √3 under Pythagorean Theorem,

In this case, $\sin 60° = \dfrac{\sqrt{3}}{2}$; and $\sin 30° = \dfrac{1}{2}$.

7

EXACT VALUES IN TRIGONOMETRY

> *For 60°, chord is 1 and supplementary chord is $\sqrt{3}$.*

Alternatively, we can take half of double angle chord for 30° in a circle. Double of 30° is 60°, which has chord of 1. Halving the angle and chord both, Sin 30°= 1/2.

Sin 18°, Sin 36° and Sin 54°

In the following triangle of 36° - 72° -72°, let's sides are a, a and 1. Bisecting one of the 72°, it forms two isosceles triangles having 36° - 72° -72° and 36° - 36° -108°. Since the ratio of sides of two similar triangle is same, taking two 36° - 72° -72° triangles (one outer and another inner); sides ratios is 1 : (a-1)= a:1 with simple algebraic form of $a^2-a=1$. Solving, $a^2-a-1=0$,

$$a = \frac{-(-1) \pm \sqrt{(-1)^2 - 4.1.(-1)}}{2.1} = \frac{1 \pm \sqrt{5}}{2}.$$

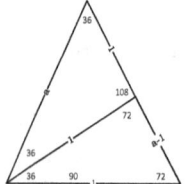

Taking positive length, value of 'a' is $\frac{1+\sqrt{5}}{2}$. Bisecting the angle having 108°, we can construct two right-angled isosceles triangles of 36° - 54° -90° having sides as 1, a/2 and $\frac{1}{2}\sqrt{4-a^2}$ using formula from Guru Pythagoras.

Taking the value of 'a' as above, the sides are 1, $\frac{1+\sqrt{5}}{4}$, and $\frac{1}{4}\sqrt{10-2\sqrt{5}}$. Form trigonometric ratios for Sine,

Sin 36°= $\frac{1}{4}\sqrt{10-2\sqrt{5}}/1 = \frac{1}{4}\sqrt{10-2\sqrt{5}}$.

Sin 54°= $\frac{1+\sqrt{5}}{4}/1 = \frac{1+\sqrt{5}}{4}$

Similarly, bisecting 36° in 36° - 72° -72° and solving for a, Sin $18° = \frac{\sqrt{5}-1}{4}$.

Alternatively, we can take half of double angle chord for 54° in a circle. Double of this makes pentagon. As per Guru Euclid, its side is $\frac{1+\sqrt{5}}{2}$. Halving the angle and chord both, Sin $54°= \frac{1+\sqrt{5}}{4}$.

In fact, geometrically, we can find the value of Sin 0°, Sin 30° and Sin 90° easily. For remaining angles, we used geo-algebraic method. So, Sin 60°, Sin 18°, Sin 36°, Sin 54°, Sin 72° have based on geo-algebraic method.

1.4 Chord and Supplementary Chord

The line that passes through two points of circumference of a circle is known as chord. Chord of an angle at the center and its acceleration factor of supplementary chord are perpendicular each other. Therefore, sum of squares of chord and its supplementary chord is 4. In this book, chord of an angle is denoted as 'a' and its supplementary chord is denoted as 'b'. In the circle in next page,

∠AOB=A⁰,

Length AB is chord for ∠A=a,

Length BD is supplementary chord of for ∠A=b,

There are certain relations regarding both chords, a and b. Based on Guru Pythagoras's theorem for perpendicular chord (a) and supplementary chord (b), $b = \sqrt{4 - a^2}$.

EXACT VALUES IN TRIGONOMETRY

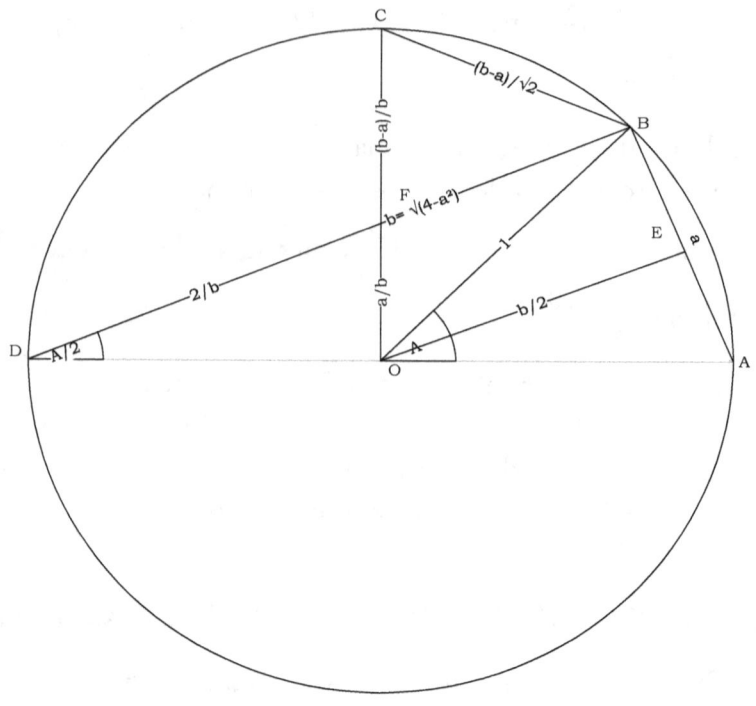

Here are some characteristics of chord (a) and supplementary chord (b) of an angle:

> T-01. *Supplementary chord (b) is acceleration factor for chord (a) of double angle.*
>
> Therefore, chord of double angle is product of 'ab'.
>
> T-02. *Apothem of a chord (line from center to mid-point of chord) will be $b/2$.*
>
> In above circle, apothem for chord AB or ∠A is OE, which is half of length of b.
>
> T-03. *Half of square of chord will be Versine of an angle.*
>
> **Versin $A = a^2/2$** and *Cos A = 1 - Versin A*

T-04. *Supplementary chord intersects the quadrant line from center at length of a/b from center. The length from that point to next point of circle is $2/b$.*

In above circle, BD intersect OC at F; then $OF = a/b$ and $BF = 2/b$.

T-05. *Chord of complementary angle of an angle is $(b-a)/\sqrt{2}$. Its supplementary will be $\sqrt{2-ab}$.*

T-06. *Chord of supplementary angle of an angle is just reversal of chord of that angle.*

$$\text{Chord } (180 - A) = b.$$

Moreover, its supplementary chord = a.

1.4.1 Half angle Chords

T-07. *Square root of the difference of 2 and supplementary chord of that angle is chord of half angle.*

T-08. *Square root of the sum of 2 and supplementary chord of that angle is chord of complementary of half angle.*

$$\text{Chord } A/2 = \sqrt{2-b}$$

$$\text{Chord } (90 - A/2) = \sqrt{2+b}$$

In fact, chord of half angle is Chord $A/2^n =$

$$\sqrt{2 - \sqrt{2 - \sqrt{\ldots 2^n - \sqrt{2 - \sqrt{4-a^2}}}}}$$ in general.

EXACT VALUES IN TRIGONOMETRY

Alternatively, it can be written as Chord $A/2^n$

$$= \sqrt{2 - \sqrt{2 - \sqrt{\ldots 2^n - \sqrt{2 - b}}}}.$$ Chord of complementary shall

be $\sqrt{2 + \sqrt{2 + \sqrt{\ldots 2^n + \sqrt{2 + b}}}}$.

Let us take few examples for this method. Chord (a) or supplementary chord (b) in half-series is always in one more nested radicals of 2 in each half.

Taking chord 1, $\sqrt{3}$ representing $60°$; where $a = 1$ and $b = \sqrt{3}$. Half-angle is $30°$, so, chord for $30°$ is $\sqrt{2 - \sqrt{3}}$. In the same line, chord for:

A	Chord (a)	$b = \sqrt{(4-a^2)}$	A/2	Crd. A/2
60	1	$\sqrt{3}$	30	$\sqrt{2-\sqrt{3}}$
30	$\sqrt{2-\sqrt{3}}$	$\sqrt{2+\sqrt{3}}$	15	$\sqrt{2-\sqrt{2+\sqrt{3}}}$
15	$\sqrt{2-\sqrt{2+\sqrt{3}}}$	$\sqrt{2+\sqrt{2+\sqrt{3}}}$	7.5	$\sqrt{2-\sqrt{2+\sqrt{2+\sqrt{3}}}}$
7.5	$\sqrt{2-\sqrt{2+\sqrt{2+\sqrt{3}}}}$	$\sqrt{2+\sqrt{2+\sqrt{2+\sqrt{3}}}}$	3.75	$\sqrt{2-\sqrt{2+\sqrt{2+\sqrt{2+\sqrt{3}}}}}$
…	…	…	…	…
36	$(\sqrt{5}-1)/2$	$\sqrt{[10+2\sqrt{5}]}/2$	18	$\sqrt{2-[\sqrt{10+2\sqrt{5}}]/2}$
18	$\sqrt{2-[\sqrt{10+2\sqrt{5}}]/2}$	$\sqrt{2+[\sqrt{10+2\sqrt{5}}]/2}$	9	$\sqrt{2-\sqrt{2+[\sqrt{10+2\sqrt{5}}]/2}}$
……	……	……	..	……

This process may be extended upto desired level to the user.

Had you care the pattern of chord (a) and supplementary chord (b), this is just single-sign difference.

TRIGONOMETRIC RATIOS

There is another method to find the half-angle chord:

T-09. *Ratio of chord of an angle to square root of the sum of 2 and supplementary chord of that angle is chord of half angle.*

If both chords of any angle are known, the chord for one-fourth (A/4) angle will be:

$$Crd\ \frac{A}{4} = \frac{a}{\sqrt{2+b}}$$

Example is Crd 60° (1, √3), Crd 30°= 1/√(2 +√3). Half of this is Sin 15°.

Half-angle chord can be calculated varieties of ways. If we see the Angle-Rewritten method of Precise-Rewritten method, each case, we computed half-angle chord. Changing the first sign of the half-chord is chord of complementary angle.

1.4.2 Double angle Chords

There is relation between chord of an angle and its double angle as:

T-10. *Chord of double angle is the product of chord and supplementary chord of that angle.*

$$\textit{Chord } 2A = ab$$

This formula is based on T-14 on page 14. Double angle chord is just 'b' times 'a' of an angle. Therefore, supplementary chord has named as '*acceleration factor*' in this book.

T-11. *Supplementary chord of double angle is the difference of 2 and square of chord of that angle.*

$$\textit{Supplementary Chord } 2A = 2 - a^2$$

1.4.3 Triple angle Chords

T-12. *Difference between 3 times of the chord with cube of same chord is the chord of triple angle.*

$$Crd\ 3A = 3a - a^3$$

For example, Crd 60° is 1, so, Crd 180°= 3.1-1³= 2.

T-13. *Difference between the ratio of chord and supplementary chord of a triple angle is 2.*

$$\frac{a}{x} - \frac{b}{y} = 2$$

$$\frac{b}{x} + 1 = b\ of\ mid-angle, where\ b$$

$$= \sqrt{2+x}$$

Where, a, b and x, y are chord and supplementary chord of angle A and A/3.

1.4.4 Addition and subtraction of Chords

T-14. *Chord for sum of angles is half of cross-multiplication of chords and supplementary chords.*

$$Chord\ (A_1 + A_2) = (b_1 a_2 + a_1 b_2)/2$$

Compute supplementary chord (b) using $\sqrt{4-a^2}$. In case both angles are same, the relation of double angled chord will be:

$$Chord\ 2A = ab$$

T-15. *Chord for difference of angles is half of cross-multiplication of chords and supplementary chords.*

$$Chord\ (A_1 - A_2) = (b_1 a_2 - a_1 b_2)/2$$

Compute supplementary chord (b) using $\sqrt{4-a^2}$.

Chord of 60° are 1, $\sqrt{3}$ and chord of 36° are $(\sqrt{5}+1)/2$, $\sqrt{10+2\sqrt{5}}/2$. Using addition formula, chord 96° is $(\sqrt{10+2\sqrt{5}} + (\sqrt{15}-\sqrt{3}))/4$. Using subtraction, chord 24° is $(\sqrt{10+2\sqrt{5}} - (\sqrt{15}-\sqrt{3}))/4$. Furthermore, chord 96° and chord 84° is same chord; and supplementary of 24° is chord 66°. Using half-angle, complementary of half-angle chord formulae (T-07 and T-08) and double angle chord (T-10), we can determine chord for 48°, 42°, 24°, 66°, 12°, 78°, 6°, 84°, 3°, 87°, 21°, 69°, 33°, 57°, 39°, 51°. Taking further calculations from half-angles from 60° or from 36° and sum and difference with 3° or other angles, chords of all integer angles divisible by 3 will obtain.

Readers may confuse here for the reason of calculating these chords. Remember half of chord 96° is Sin 48° or half of chord 24° is Sin 12°.

1.4.5 Calculated Chords

From the chord (a) and supplementary chord (b), we may know the length of Sine of angle without using any table or calculators. We can compute chord of angles: using Addition and subtraction of (pg. 14) above, all the angles in degrees divisible by 3 can be calculated. Hence, our inventory of available chords is 30 within the first quadrant[5] taking integer angles. In case an angle having decimal number, angle in degrees that is divisible by 3 as well as last 2-digits after decimal are divisible by 25, they have exact value of chord.

---

[5] Even angles more than 90° has widely use in this book, but the core idea has concentrated upto 90° only. Few cases, it limits upto 45°, leaving complementary angles.

EXACT VALUES IN TRIGONOMETRY

Using Chord Relations Half angle Chords (pg. 11), we can calculate half-series of each of angles divisible by 3. The number of angles having calculated chord shall have infinitive. Double angle Chords (pg. 13), Triple angle Chords (pg. 14) will not contribute only few new chords.

From above known chords, we can calculate exact value of trigonometric ratios of that angle. This is why, we can compute exact value which is divisible by 3 (taking all digits in a number having decimal, if it contains decimal) and further last 2 digits after the decimal divisible by 25. For example, Sin 69° is divisible by 3 and last two digits after decimal of 69.00 is divisible by 25, so, it has exact value of Sine. For Sin 1.875°, this is divisible by 3 as well as last two digits are also divisible by 25; it has exact value. For Sin 0.5625°, it is divisible by 3 as well as last two digits are divisible by 25 and hence have exact value. If an angle is not perfectly divisible by 3, we can compute exact value of Sine to the angle nearest to that angle having last divisible 25 after decimal. For example, for 1°, nearest number divisible by 0.999999..975 or 1.0000...05, for 40, nearest would be 39.999..975 or 40.000..005 will have exact value of Sine. Considering the *appropriate accuracy*, we may use the exact value for any angle whether divisible by 3 or not (for basis and details, see Chapter 5.

From chord (a), supplementary chord (b) and their product (ab), we can calculate any trigonometric relations (see the list of formulae on page 2) using following methods:

    ARC-LINE METHOD (Pg. 21)

    SERIES-METHOD (Pg. 29)

    ANGLE-REWRITTEN METHOD (Pg. 47)

    PRECISE-REWRITTEN METHOD (pg. 65)

TRIGONOMETRIC RATIOS

1.5 Sine Square

Let us develop few of sine from single picture as below:

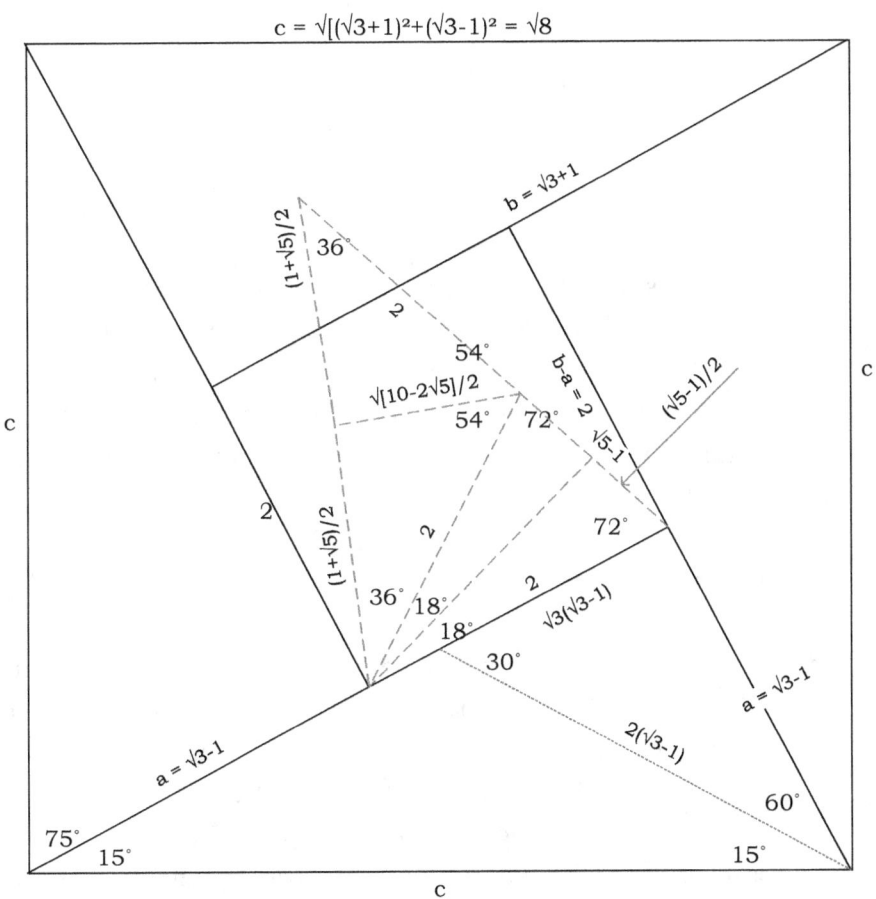

Interpretation and proofs on the picture:

1. The main square had developed by Guru Bhaskara in 11th Century. This square describes the proof of Guru Pythagoras theorem of sum of squares of sides of a right angle triangle is equal to the square of its hypotenuse.

   There are 4 right-angled triangles and a small square within the main square having side 'c'. Sides of triangle has

17

assumed as 'a' and b whereas hypotenuse is 'c'. Side of inner square is 'b-a'.

Area of outer square is c^2. In the same time, area of same square is the sum of area of 4 right-angled triangles and inner-square. Using area formula:

Area of 4- triangles= 4. ½.a.b =2ab

Area of inner-square= $(b-a)^2 = a^2+b^2 - 2ab$

∴ Area of outer-square, $c^2 = a^2+b^2$

2. Let us assume the angles made by 'a' and 'c' is 15^0, so angle formed by b and c will be 75^0. Deliberately taking a= $\sqrt{3}-1$, b=$\sqrt{3}+1$, c becomes $\sqrt{8}$ using above formula and side of inner square (b-a) will be 2. Above formula permits our values. In this case, we can find the value of Sin 15^0 and Sin 75^0.
Sin 15^0 = $(\sqrt{3}-1)/\sqrt{8}$ = $(\sqrt{3}-1)/2\sqrt{2}$
Sin 75^0 = $(\sqrt{3}+1)/\sqrt{8}$ = $(\sqrt{3}+1)/2\sqrt{2}$

3. If we take the diagonal of inner-square, its length will be $\sqrt{8}$ based on above formula (this has not been drawn in above picture) and the angle will be 45^0 (diagonal bisect the angle in square). Hence Sin 45^0 will be $2/\sqrt{8}=1/\sqrt{2}$.

4. In the dotted triangle, outer-triangle is 36^0-72^0-72^0 with base of 2 and 2-unknown sides (assuming each side is 's'). Bisecting one of 72^0, each part becomes 36^0. Two isosceles triangles having 36^0 -36^0 -108^0 and 36^0-72^0-72^0 formed. Now, we have two similar triangles of 36^0-72^0-72^0. Using the properties of similar triangle have same ratios with the corresponding sides, $\frac{s}{2} = \frac{2}{s-2}$. Therefore, we obtain, $s^2-2s=4$. Using solution of quadratic equation $s = 1 \pm \sqrt{5}$. However, we know the length of side shall be a positive parameter, so, we have single candidate of s=$1+\sqrt{5}$.

TRIGONOMETRIC RATIOS

In the next step of calculation, bisect the angles of 108^0 and 36^0 as above; resulting angles will be 54^0 and 18^0. The corresponding sides will be $(\sqrt{5}+1)/2$ and $(\sqrt{5}-1)/2$ respectively. Now we can compute easily,

Sin $18^0 = [(\sqrt{5}-1)/2]/2 = (\sqrt{5}-1)/4$
Sin $36^0 = [\sqrt{(10-2\sqrt{5})}/2]/2 = [\sqrt{(10-2\sqrt{5})}/4$
Sin $54^0 = [(\sqrt{5}+1)/2]/2 = (\sqrt{5}+1)/4$
Sin $72^0 = [\sqrt{(10+2\sqrt{5})}/2]/2 = [\sqrt{(10+2\sqrt{5})}/4$

5. What's about 30^0 and 60^0? They are shown in the picture; but we know better way than that shown above. Its purpose is to express the variety of way of presentation only. Similar presentation would be possible if we replaced the 15^0 and 75^0 by 18^0 and 72^0 in the above picture.

1.6 Sine Scale

We use semi-lunar shaped protector to measure the angle, normally available in degrees and runs from $0°$ to $180°$. The diameter line and degree marks has shown by default by the manufacturer.

Now, just make a parallel line joining each degree with its supplementary angle and mark in the middle. These marks are marks for Sine Scale starting 0 in the center of circle. The diameter in the Sine-Scale is recommended is 2. Based on the scale from protector, as mentioned above, a simple straight ruler may be construct in the perpendicular line as Sine-Scale. If the line is marked in diameter (starting 0 in center) that will measure Cosine of any angle fall down to the diameter. To measure small scales, as similar as in other straight edge having micro-measurement, Sine-line with measuring tools with micro-unit may manufacture.

EXACT VALUES IN TRIGONOMETRY

Just a Circle
Concept of half-angles and centrals

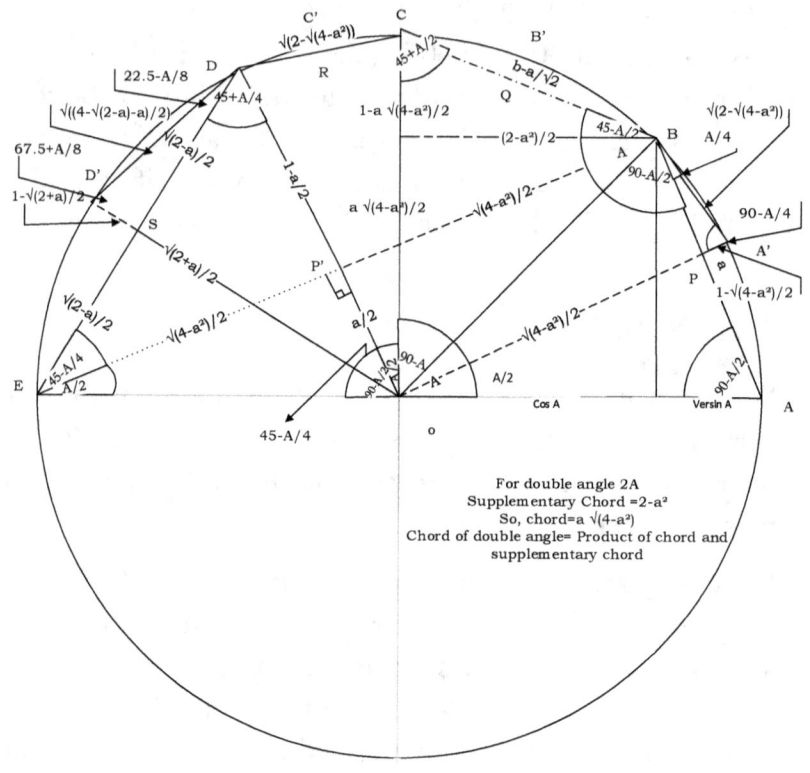

Chapter 2
ARC-LINE METHOD

KEY WORDS – arc-line, chord, acceleration factor, power-base, appropriate accuracy.

2.1 Arc-Line concept

Generally a human can see 5 Km. in the horizon as a plain earth surface. A bed above the earth is a plain object. We can put an office-table in the floor or in open ground. A football ground or farming land seems plain. Quadruped (Four-leg) animals run above the earth surface without any difficulties. All these day-to-day items are straight above the earth surface, but we know the earth itself is circular sphere.

The arc of an angle is circular in nature, not a straight line, but a tiny part of arc is straight. Area of circle is determined based on straight-line concept of circumference[6]. In this chapter, we use similar concept, as the tiny part of a circular segment is straight for determination of trigonometric ratios.

---

[6] If a circle converted into a right-angled triangle, the height will be the radius (r) with circumference ($2\pi r$) as base; therefore, the area is $2\pi r^2$.

EXACT VALUES IN TRIGONOMETRY

2.1.1 Inputs

For this method, we need three generic inputs: radian measure of angle, power-base and chord formula.

- Radian measurement of angle is a general input.
- Power-base must be in 2^n pattern; where minimum value of n is 4 and recommended value is 16 or more.
- Chord of an angle in center of circle is 'a' and acceleration factor (supplementary chord (b) for same angle will be $\sqrt{4-a^2}$ and vice versa. The product of 'a' and 'b' is chord 'a' for double angle (see T-10 on page 13). Half of product of 'a' and 'b' is value of Sine of that angle (see T-16 on page 31).

Based on above three inputs, we can compute Sine of *any angle* using the simple arithmetical process.

2.1.2 Accuracy

Accuracy under Arc-Line method depends upon three parameters – level of power-base (2^n), number of digits in each figure during calculations and accuracy of π or τ itself.

If we take n=4 for power-base 2^n, the result will be acceptable for general purpose. To increase accuracy, increase the power-base 2^n (as 2^{10}, 2^{16}, 2^{20}, 2^{50}, 2^{100}, 2^{1000} etc.). Just a power-base of 2^4 gives accuracy of 9 decimal places for the small angles and 7 decimal places for the large angles within the first quadrant.

If we measure their relations, roughly, 2^{13} part of earth circumference has reserved as straight for a human eye, seeing as plain earth. A Grand King-sized bed is 2^{24} part of earth, whereas office table is 2^{25} part, a football ground is 2^{20} part, and human foot is 2^{32} part of earth circumference.

ARC-LINE METHOD

In more general, for angle base of 1° for earth as a circle:

- Power-base 2^{12} catches the circular segment below a commercial building, because, nearly it gives residential house as straight; it catches the gate or yard of residential house is circular comparing to that residential-building. [In this case, calculation-steps will be 13 for 1°].
- Power-base 2^{14} catches the circular segment below a small-sized residential building, because, nearly it gives a small bed-room as straight; it gives adjoining rooms in circular part of circular earth surface. [In this case, calculation-steps will be 15 for 1°].
- Power-base 2^{16} catches the circular segment below a bed, because, nearly it gives an office-table as straight; it gives a bed and adjoining table in different circular part of earth surface. [In this case, calculation-steps will be 17 for 1°].
- Power-base 2^{40} catches around 1" of surface of Sun as straight. A bullet thrown towards the surface of the Sun from earth, hits in the exact point as we prescribed so far. [In this case, calculation-steps will be 41 for 1°].

Accuracy (*appropriate accuracy*) also depends upon number of digits handled during each process of calculation. If we define the appropriate accuracy of 15-digits after decimal point, we must use, *at least,* 15-dgits after decimal in each steps of calculation.

Need not to say, if the angle is based on π or τ, accuracy of value of trigonometric ratios depends upon the accuracy of value of π or τ. For example, if we take value of π as $\sqrt{10}$, of course, we may guess the extreme low accuracy than 22/7 or 355/113 or other modern value of π.

Though discussed three parameters of accuracy, in the example below, we compromised the accuracy of π, we limited our calculation in 14-digits after decimal.

EXACT VALUES IN TRIGONOMETRY

2.1.3 Mathematical process

In this method, one need to perform just simple arithmetical process for *n+1* steps for power of 2^n. In case, we take the base as 2^4, the calculation process will limit to 5 steps only. If we take the base as 2^{16}, the calculation process will limit to 17 steps. Incredible point is that, we have 'a' and 'b' in each step, therefore every step gives trigonometric ratios of each angle in the 2^n series with same *appropriate accuracy*.

Determination of trigonometric ratios of an angle or chord-length or its acceleration factor of supplementary chord has five steps of arithmetic calculation.

> Let us, explain the method of determination of Sin 64 degrees (71.11111… grades) as an example:

Firstly, convert the angle into radian with appropriate decimal places of accuracy.

> In our case it is $A = 1.11701072127637^c$. In case, the angle itself is in radian, no conversion requires.

Secondly, take the appropriate power-base of 2 as 2^4, 2^{10}, 2^{16}, 2^{50}, or 2^{100}, or higher. Calculation steps will be n+1.

> Let us take the power-base of 2^{16} in our example. We need to multiply 17 times to arrive Sine of 64°.

Thirdly, find the arc length 2^nth part of given angle. For this, divide the angle in radian by power-base. This will be the first chord.

First Chord (θ) = $A^c / 2^n$

> In our example, $1.11701072127637 / 2^{16}$ is the first chord. An angle θ has used here is just for showing the flow of calculation.

ARC-LINE METHOD

Fourthly, take the chord (a) and find the acceleration factor as supplementary chord $b = \sqrt{4 - a^2}$.

In our example, acceleration factor of first chord is $b = \sqrt{4 - (1.11701072127637/2^{16})^2}$.

Fifthly, multiply both 'a' and 'b', which shall be the chord (a) of double of angle, i.e. chord 2θ.

From this chord 'a' of 2θ, find acceleration factor 'b' using same formula in fourth step above. Product of 'a' and 'b' will be the chord of double angle, i.e. 4θ. In the meantime, halving this product of 'a' and 'b' is Sine of that angle.

Chord 2θ = a×b (of θ see T-10 on page 13)

Sin θ = a×b/2 (of θ see T-16 on page 31)

Repeat the process in fourth and fifth steps for n+1 times to arrive to the given angle (i.e. 2^n θ that is the given angle A). Dividing the product of 'a' and 'b' by diameter 2 in each step is Sine of that angle. Hence, first time, it gives the value of Sin θ; second time Sin 2θ; third time Sin 4θ, fourth time Sin 8θ and so on. In calculation during the n+1 time (in our case, 17th time), it gives Sin 2^n θ, which is Sin A or Sin 64° in our case.

Degrees	Radian (A)	Chord (a)
0.00098	0.00001704423097650710	0.00001704423097650710 0
0.00195	0.00003408846195301420	0.00003408846195177630 0
0.00391	0.00006817692390602840	0.00006817692389364980 0
0.00781	0.00013635384781205700	0.00013635384770807600 0
0.01563	0.00027270769562411400	0.00027270769478236700 0
0.03125	0.00054541539124822700	0.00054541538449445200 0
0.06250	0.00109083078249645000	0.00109083072842664000 0
0.12500	0.00218166156499291000	0.00218166113235520000 0

25

EXACT VALUES IN TRIGONOMETRY

0.25000	0.004363323129985820000	0.004363319668726350000
0.50000	0.008726646259971640000	0.008726618569598760000
1	0.017453292519943300000	0.017453070996959100000
2	0.034906585039886500000	0.034904812874989400000
4	0.069813170079773100000	0.069798993405846400000
8	0.139626340159546000000	0.139512947489936000000
16	0.279252680319092000000	0.278346201923478000000
32	0.558505360638185000000	0.551274711640496000000
64	1.117010721276370000000	1.059838528477870000000

Table contd…..

Degrees	$b = \sqrt{4 - a^2}$	$Sin\ A = a \times b/2$
0.00098	1.999999999927370000	0.0000170442309758882
0.00195	1.999999999709490000	0.0000340884619468249
0.00391	1.999999998837980000	0.0000681769238540382
0.00781	1.999999995351910000	0.0001363538473911840
0.01563	1.999999981407630000	0.0002727076922472260
0.03125	1.999999925630510000	0.0005454153642133210
0.06250	1.999999702522060000	0.00109083056617760
0.12500	1.999998810088320000	0.002181659834363170
0.25000	1.999995240354700000	0.004363309284799380
0.50000	1.999980961441470000	0.008726535498479550
1	1.999923846128340000	0.017452406437494700
2	1.999695390312780000	0.034899496702923200
4	1.998781654038160000	0.069756473744968200
8	1.995128100519530000	0.139173100961739000
16	1.980536137482670000	0.275637355820248000
32	1.922523391874770000	0.529919264238937000
64	1.696096192305690000	0.898794046305093000

2.2 Trigonometric Ratios of any angle

From above method, we can calculate Sine of any integer angle or other angles, as we want to find.

For example, if we want to find the Sine of angle 16.5° with extremely high accuracy[7] of decimal places, take the power-base of 2^{100} (such high power-base gives accuracy for a giant circle like star from earth). For each calculation, take higher digits after decimal point, say 30-digits or more. In this case, our calculation process will repeat for 101 times as:

- Firstly calculate radian of 16.5° and divide that by 2^{100} to get the chord (a) of $(16.5/2^{100})°$. Find the acceleration factor $b = \sqrt{4 - a^2}$.
- Multiply 'a' and 'b'; product will be chord 'a' of $(16.5/2^{99})°$. Find the value of acceleration factor $b = \sqrt{4 - a^2}$.
- Repeat the process for 101 times; the half of 'a' and 'b' of 101st time is Sin 16.5°.

What to do, if we need value for Sin 33° or Sin 66°? Nothing special, product of 'a' and 'b' of 16.5° will be the chord 'a' for 33°. Product of 'a' and 'b' of 33° will be the chord of 66°. Half of the product of chord (a) and supplementary chord (b) is Sine of that angle in all cases.

In the above process, we obtained both 'a' and 'b' in each step. Therefore, we can calculate all trigonometric ratios (Cos A, Tan A etc.) using formulae for each trigonometric relation given on page 2.

---

[7] Indicating one point is good here: Except value of Sine of 0°, 30°, 90° and 180°, all the Sine of angles are irrational number in nature.

2.3 Angle-based Formula

Determination of Sine of angle may be possible using lower angle base of multiple of 2 from the given angle and lower power-base. For example, we know, $64° = 1° \times 2^6$. Then why need to take $64°$ in our calculation, take just $1°$ with power-base 2^4. The result can increase for $64°$ easily. For other cases, reduce the angle as small as possible.

For any angles β is the radian measure of the given angle.

$$\text{Sin } \beta = \frac{1}{2}[2 - (16\beta^2 - 336\beta^4 + 1072\beta^6 - 760\beta^8 + 352\beta^{10} - 104\beta^{12} + 16\beta^{14} - \beta^{16})] \times$$

$$\sqrt{4 - \left[2 - \left(\begin{array}{c}16\beta^2 - 336\beta^4 + 1072\beta^6 - 760\beta^8 \\ + 352\beta^{10} - 104\beta^{12} + 16\beta^{14} - \beta^{16}\end{array}\right)\right]^2}.$$

This gives accuracy of 9-decimal digits for small angles as β and 7-digits for large angles as β. Of course, the formula itself seems irritating for calculation.

Based on above, we may generalize the formula for Sine of any angle (A) in radian measurement as:

$\text{Crd } A/2^n \qquad = A/2^n$

$\text{Crd } A/2^{n-1} \quad = 2 \text{ Sin } A/2^n \quad = A/2^n * \sqrt{4 - (A/2^n)^2}$

$\text{Crd } A/2^{n-2} \quad = 2 \text{ Sin } A/2^{n-1} = \text{Crd } A/2^{n-1} * \sqrt{4 - [\text{Crd } A/2^{n-1}]^2}$

....

$\text{Crd } A/2^0 \quad = 2 \text{ Sin } A/2^1 \quad = \text{Crd } A/2^1 * \sqrt{4 - [\text{Crd } A/2^1]^2}$

$\text{Crd } A/2^{-1} = 2 \text{ Sin } A \quad = \text{Crd } A/2^0 * \sqrt{4 - [\text{Crd } A/2^0]^2}$

$$\text{Sin } A = \frac{\text{Crd } A * \sqrt{4 - [\text{Crd } A]^2}}{2}$$

Chapter 3
SERIES-METHOD

KEY WORDS –chord, acceleration factor, supplementary chord, known-chord, calculated-chord, half-chord, power-base.

3.1 Exact Sine from Series Method

"Half of the chord of any angle is Sine of half of that angle" is well known geometric formula for finding Sine of an angle since than 2,000 years. Greek and Hindu mathematicians separately documented this concept as chord and Jyardha.

Why half of the chord is Sine of half angle, has simple solution. Central angle is A and angle in circumference in the left is A/2 (being half of central angle). Using the definition of Sine as ratio of perpendicular to the hypotenuse, chord of A is the opposite side to A/2 and its hypotenuse is the diameter having length of 2. So, ratio is chord/2 is the Sine of A/2.

EXACT VALUES IN TRIGONOMETRY

Under this method, if length of chord is known, Sine of half angle is half of that chord. For example, chord of 60° is 1.

Length of chord of 60° is 1; half of both angle and chord 30⁰ and 0.5. Therefore, Sin 30° = 0.5

In the same time, we can easily find the supplementary angle and supplementary chord. Using the same theorem, we can calculate the Sine of half of supplementary angle.

In above case, supplementary angle is 120° and length of supplementary chord is $\sqrt{3}$. Halving both, Sin 60° is $\sqrt{3}/2$.

If we carefully observe the value of Sine of any angle that shall be half-chord framed by double of that angle. In case, Sine of angle has calculated by using other method that also would follow this half-chord theory. Thermos discussed below are usable in Sine of an angle from its chord (a) and supplementary chord, $b = \sqrt{4 - a^2}$.

Under Series-method, we can find the exact values of trigonometric relations (in this book only Sine of an angle has dealt with) in the series of angle as $A/2^n$. For example, we determine the exact value in series like A/1-series, A/2-series, A/4-series, and so on. Since these are exact value, we need not worried on accuracy. However, for decimal representation, we should consider *appropriate accuracy*.

3.1.1 Angle and Complementary-angle Series

In the Series method, we segregates the main angle into a series based on $A/2^n$. Therefore, we can compute the exact value of Sine of A-series, A/2-series, A/4-series, A/8-Series, ..., A/n-series. For the purpose of Series-method, A is the angle whose chord (a) and supplementary chord (b) is known. For this purpose, chord of 90°, 60° and 36° is known to us; so A may be any of these angles.

SERIES METHOD

Value of Sine in A/1-series

Sin (90°±A)

Each A/n-series contains two routes as right-angle-route (90°) and given angle-route (A). In A/1-series, central-angles are 90°/1±A/1= 90°±A. However, 90°-A is complementary of A, so, there is only one central-angle 90°+A=A. Being the first series, there is no earlier central-angles. Therefore, A/1-series has two wings of angles as Sin A and complementary angle Sin (90-A).

> T-16. *Half of product of chord and its supplementary chord is Sine of that angle.*
>
> *Sin A= a b/2*

Alternatively, Sin A= $a \cdot \sqrt{4 - a^2}$;

Crd 60° is 1, √3. So, Sin 60°= 1×√3/2= √3/2.

Crd 90° is √2, √2. So, Sin 90°= √2×√2/2= 1.

> T-17. *Half of difference between 2 and square of chord of an angle is Sine of complementary of that angle.*
>
> *Sin (90-A) = (2- a^2) /2*   *[for A<90°]*

From Crd 60° (1, √3), Sin 30°= *(2- 1^2) /2*= 1/2. From supplementary chord, Sin 60°= √3/2.

From Crd 90° (√2, √2), Sin 0°= *(2- $\sqrt{2}^2$) /2*= 0 and Sin 45° =√2/2.

31

EXACT VALUES IN TRIGONOMETRY

Value of Sine in A/2-series

Sin (90°±A/2) and Sin (45°±A/2)

Each A/n-series contains two routes as right-angle-route (90°) and given angle-route (A). In A/2-series, there are four angles as:

- Central-angles are $90°/2 \pm A/2 = 45° \pm A/2$. Since both 90/n and A/n has same value of n-dependency is 'ab'.
- Earlier central-angles for A/2-series is central angles of A/1-series- which is A; halving A/2, dependency is 'a'.
- Remaining and complementary angles in A/2-series is 90°-A/2, dependency is b.

We know the well-known Sine formula of half-chord since last 2000 years as:

T-18. *Half of a chord is Sine of half of that angle.*

$$Sin\ A/2 = a/2$$

In the next hand, supplementary angle and chord has similar characteristics as above. So, half of supplementary chord is Sine of half of supplementary angle. Fortunately, this angle is complementary angle of half of that angle. So,

T-19. *Half of supplementary chord is Sine of complementary angle of half of that angle.*

$$Sin\ [(180°-A)/2] = \sqrt{4-a^2}/2 = b/2$$
$$Sin\ (90°-A/2) = b/2$$

T-20. *Half of square root of difference between 2 and product of chord and supplementary chord is Sine of 45° less half of that angle.*

$$Sin\ (45°-A/2) = \sqrt{2-ab}/2$$

This theorem may be re-written as:

SERIES METHOD

T-21. *Half of square root of difference between 2 and product of chord and supplementary chord is Sine of half of complementary angle of that angle.*

$$Sin\ [(90^0-A)/2] = \sqrt{2-ab}/2$$

T-22. *Half of square root of sum of 2 and product of chord and supplementary chord is Sine of 45⁰ plus half of that angle.*

$$Sin\ (45^0+A/2) = \sqrt{2+ab}/2$$

This theorem may be re-written as:

T-23. *Product of square root of 2 and average of chords is Sine of 45⁰ plus half of that angle.*

$$Sin\ (45^0+A/2) = \frac{a+b}{2} \times \sqrt{2}.$$

This theorem may be re-written as:

T-24. *Square root of 2 times the sum of chords is Sine of 45⁰ plus half of that angle.*

$$Sin\ (45^0+A/2) = \frac{a+b}{2} \times \sqrt{2} = \sqrt{2}(a+b).$$

This theorem may be re-written as:

T-25. *Half of square root of difference between 2 and product of chord and supplementary chord is Sine of sum of that angle and half of complementary angle of that angle.*

$$Sin\ (A+(90-A)/2) = \sqrt{2+ab}/2$$

EXACT VALUES IN TRIGONOMETRY

From Crd 60° (1, √3), Sin 30° = ½; Sin 60° = √3/2; Sin 15° = √(2 −1×√3)/2 = √(2 −√3)/2; Sin 75° = √(2 +√3)/2.

From Crd 90° (√2, √2), Sin 45° = √2/2; Sin 0° = 0; Sin 90° = √(2 +√2×√2)/2 = 1.

Value of Sine in A/4-series

Sin (90°±A/4), Sin (45°±A/4), Sin (22.5°±A/4) and Sin (67.5°±A/4)

In A/4-series, there are eight (2^{n+1}) angles as:

- Last − Central has n-numbers of angles. In this case, last central is 90/4 = 22.5 written as 90°/2 ± (90°/2 ± A/2)/2 = 45° ± 22.5° ± A/4. Since last item of 90/n and A/n has same value of n- dependency is 'ab'.
- Second-last Central, for A/4-series is central angles of A/2-series− which is 45°±A/2; halving the angle only, it is 45°±A/4. Since last item of 90/n and A/n has the ratio of 2- dependency is 'a'.

SERIES METHOD

- Remaining and complementary angles in A/4-series are 90°-A/4 and A/4. Since last item of 90/n and A/n has the ratio of 4 (or 8, 16 …)- dependency is 'b'.

	2nd last-Central	Last Central	Earlier Centrals
Earlier angle	A/2	45±A/2	90 − A/2
A/4-series	45±(A/2)/2	45± (45±A/2)/2	90 ± (A/2)/2
	45±A/4	22.5±A/4	A/4
		67.5±A/4	90 − A/4
Dependency	'a'	'ab'	'b'

T-26. *Half of square root of difference between 2 and supplementary chord of an angle is Sine of one-fourth of that angle.*

$$Sin\ A/4 = \sqrt{(2-b)}\ /2$$

E-01 Let's take following unexpanded examples:

Crd 90° (√2, √2)   Sin 22.5° = $\sqrt{(2-\sqrt{2})}\ /2$

Crd 60° (1, √3)   Sin 15° = $\sqrt{(2-\sqrt{3})}\ /2$

T-27. *Half of square root of sum of 2 and supplementary chord of an angle is Sine of complementary angle of one-fourth of that angle.*

$$Sin\ (90-A/4) = \sqrt{(2+b)}\ /2$$

T-28. *Half of square root of difference between 2 and chord of an angle is Sine of difference of 45° and one-fourth of that angle.*

$$Sin\ (45°-A/4) = \sqrt{(2-a)}\ /2$$

EXACT VALUES IN TRIGONOMETRY

T-29. Half of square root of sum of 2 and chord of an angle is Sine of sum of $45°$ and one-fourth of that angle.

$$Sin(45°+A/4) = \sqrt{(2+a)}/2$$

T-30. Half of square root of sum of 2 and square root of difference of 2 and product of chord and supplementary chord is Sine of difference of $67.5°$ and one-fourth of that angle.

$$Sin(67.5°-A/4) = \sqrt{(2+\sqrt{(2-ab)})}/2$$

T-31. Half of square root of sum of 2 and square root of 2 plus product of chord and supplementary chord is Sine of sum of $67.5°$ and one-fourth of that angle.

$$Sin(67.5°+A/4) = \sqrt{(2+\sqrt{(2+ab)})}/2$$

T-32. Half of square root of difference of 2 and square root of difference of 2 and product of chord and supplementary chord is Sine of sum of $22.5°$ and one-fourth of that angle.

$$Sin(22.5°+A/4) = \sqrt{(2-\sqrt{(2-ab)})}/2$$

T-33. Half of square root of difference of 2 and square root of 2 plus product of chord and supplementary chord is Sine of difference of $22.5°$ and one-fourth of that angle.

$$Sin(22.5°-A/4) = \sqrt{(2-\sqrt{(2+ab)})}/2$$

Value of Sine of A/8 series

Each A/n-series contains two routes as right-angle-route ($90°$) and given angle-route (A). In A/8-series, there are sixteen angles as:

- Central -angles are $90°/2 \pm (90°/2 \pm (90°/2 \pm A/2)/2)/2$ = $45° \pm 22.5° \pm 11.25° \pm A/8$. Since last item of $90/n$ and A/n has same value of n- dependency is 'ab'.

SERIES METHOD

- Earlier central-angles for A/8-series is central angles of A/4-series- which is 45°±22.5°±A/4; halving the angle only, it is 45°±22.5°±A/8. Since last item of 90/n and A/n has the ratio of 2- dependency is 'a'.
- Remaining and complementary angles in A/8-series are 90°±A/8 and 45°±A/8. Since last item of 90/n and A/n has the ratio of 4 or 8, 16 …- dependency is 'b'.

	Earlier-central –angle	Central-angles	Remaining angles
Earlier angle	45±A/4	45± 22.5±A/4	90 ± A/4
A/8-series	45± 22.5±A/8	45±22.5±11.25±A/8	90 ± (A/4)/2 45±(A/4)/2
	22.5±A/8 67.5±A/8	11.25±A/8 33.75±A/8 56.25±A/8 78.75±A/8	A/8 90 – A/8 45±A/8
Dependency	'a'	'ab'	'b'

Sine for the angles in A/8-series are as follows[8]:

1. $Sin\ A/8$ = $\sqrt{[2 - \sqrt{(2 + b)}]}/2$
2. $Sin\ (90°-A/8)$ = $\sqrt{[2 + \sqrt{(2 + b)}]}/2$
3. $Sin\ (45°+A/8)$ = $\sqrt{[2 + \sqrt{(2 - b)}]}/2$
4. $Sin\ (45°-A/8)$ = $\sqrt{[2 - \sqrt{(2 - b)}]}/2$
5. $Sin\ (22.5°-A/8)$ = $\sqrt{[2 - \sqrt{(2 + a)}]}/2$

---

[8] To make the presentation easy, all the closing brackets are collapsed by single bracket ']' in this Chapter.

EXACT VALUES IN TRIGONOMETRY

6. $Sin\ (22.5°+A/8)$ $=\sqrt{[2-\sqrt{(2-a)}]/2}$
7. $Sin\ (67.5°+A/8)$ $=\sqrt{[2+\sqrt{(2+a)}]/2}$
8. $Sin\ (67.5°-A/8)$ $=\sqrt{[2+\sqrt{(2-a)}]/2}$
9. $Sin\ (56.25°+A/8)$ $=\sqrt{[2+\sqrt{(2-\sqrt{(2-ab)}}]/2}$
10. $Sin\ (56.25°-A/8)$ $=\sqrt{[2+\sqrt{(2-\sqrt{(2+ab)}}]/2}$
11. $Sin\ (33.75°-A/8)$ $=\sqrt{[2-\sqrt{(2-\sqrt{(2-ab)}}]/2}$
12. $Sin\ (33.75°+A/8)$ $=\sqrt{[2-\sqrt{(2-\sqrt{(2+ab)}}]/2}$
13. $Sin\ (78.75°-A/8)$ $=\sqrt{[2+\sqrt{(2+\sqrt{(2-ab)}}]/2}$
14. $Sin\ (78.75°+A/8)$ $=\sqrt{[2+\sqrt{(2+\sqrt{(2+ab)}}]/2}$
15. $Sin\ (11.25°-A/8)$ $=\sqrt{[2-\sqrt{(2+\sqrt{(2+ab)}}]/2}$
16. $Sin\ (11.25°+A/8)$ $=\sqrt{[2-\sqrt{(2+\sqrt{(2-ab)}}]/2}$

Above formulae are based on chord (half-chord formula). These angles are calculative based on Sine of earlier angle itself. For example, $Sin\ (45°\pm A/8) = \sqrt{(1\pm Sin\ A/4)}$ or $Sin\ (22.5°\pm A/8) = \sqrt{[1\pm Sin\ (45°-A/4)]}$. This formula is equally applicable in other set of above or below angles.

Value of Sine of A/16 series

In A/16-series, there are thirty-two angles as:

- Central -angles are $90°/2\pm (90°/2\pm(90°/2\pm(90°/2\pm A/2)/2)/2)/2 = 45°\pm 22.5°\pm 11.25°\pm 5.625°\pm A/16$. Since last item of 90/n and A/n has same value of n- dependency is 'ab'.
- Earlier central-angles for A/16-series is central angles of A/8-series- which is $45°\pm 22.5°\pm 11.25°\pm A/8$; halving the angle only, it is $45°\pm 22.5°\pm 11.25°\pm A/16$. Since last item of 90/n and A/n has the ratio of 2-dependency is 'a'.
- Remaining and complementary angles in A/16-series are $90°\pm A/16$, $45°\pm A/16$ and $45°\pm 22.5°\pm A/16$. Since

SERIES METHOD

last item of 90/n and A/n has the ratio of 4 or 8, 16 ...- dependency is 'b'.

Sine of A/16-series angles are based on following formulae:

1. Sin A/16   =√(2 -√(2 + √(2 +b]/2
2. Sin (90-A/16)   =√(2 + √(2 + √(2 +b]/2
3. Sin (45+A/16)   =√(2 +√(2 -√(2 +b]/2
4. Sin (45-A/16)   =√(2 - √(2 -√(2 +b]/2
5. Sin (22.50 - A/16)   =√(2 -√(2 + √(2 -b]/2
6. Sin (22.50+A/16)   =√(2 -√(2 -√(2 -b]/2
7. Sin (67.50+A/16)   =√(2 + √(2 +√(2 -b]/2
8. Sin (67.50-A/16)   =√(2 + √(2 -√(2 -b]/2
9. Sin (56.25+A/16)   =√(2 + √(2 -√(2 -a]/2
10. Sin (56.25-A/16)   =√(2 + √(2 -√(2 +a]/2
11. Sin (33.75-A/16)   =√(2 -√(2 -√(2 -a]/2
12. Sin (33.75+A/16)   =√(2 -√(2 -√(2 +a]/2
13. Sin (78.75-A/16)   =√(2 + √(2 + √(2 -a]/2
14. Sin (78.75+A/16)   =√(2 + √(2 + √(2 +a]/2
15. Sin (11.25-A/16)   =√(2 -√(2 + √(2 -a]/2
16. Sin (11.25+A/16)   =√(2 -√(2 + √(2 +a]/2

17. Sin (39.375+A/16)   =√(2 -√(2 -√(2 + √(2 + ab]/2
18. Sin (39.375-A/16)   =√(2 -√(2 -√(2 + √(2 - ab]/2
19. Sin (28.125+A/16)   =√(2 -√(2 -√(2 -√(2 - ab]/2
20. Sin (28.125-A/16)   =√(2 -√(2 -√(2 -√(2 + ab]/2
21. Sin (16.875+A/16)   =√(2 -√(2 + √(2 -√(2 + ab]/2
22. Sin (16.875-A/16)   =√(2 -√(2 + √(2 -√(2 - ab]/2
23. Sin (50.625+A/16)   =√(2 + √(2 -√(2 + √(2 - ab]/2

EXACT VALUES IN TRIGONOMETRY

24. $\sin(50.625 - A/16) = \sqrt{2 + \sqrt{2 - \sqrt{2 + \sqrt{2 + ab}}}}/2$
25. $\sin(61.875 + A/16) = \sqrt{2 + \sqrt{2 - \sqrt{2 - \sqrt{2 + ab}}}}/2$
26. $\sin(61.875 - A/16) = \sqrt{2 + \sqrt{2 - \sqrt{2 - \sqrt{2 - ab}}}}/2$
27. $\sin(73.125 - A/16) = \sqrt{2 + \sqrt{2 + \sqrt{2 - \sqrt{2 + ab}}}}/2$
28. $\sin(73.125 + A/16) = \sqrt{2 + \sqrt{2 + \sqrt{2 - \sqrt{2 - ab}}}}/2$
29. $\sin(5.625 - A/16) = \sqrt{2 - \sqrt{2 + \sqrt{2 + \sqrt{2 + ab}}}}/2$
30. $\sin(5.625 + A/16) = \sqrt{2 - \sqrt{2 + \sqrt{2 + \sqrt{2 - ab}}}}/2$
31. $\sin(84.375 - A/16) = \sqrt{2 + \sqrt{2 + \sqrt{2 + \sqrt{2 - ab}}}}/2$
32. $\sin(84.375 + A/16) = \sqrt{2 + \sqrt{2 + \sqrt{2 + \sqrt{2 + ab}}}}/2$

3.1.2 Sine of $A/2^n$-Series

If we care the pattern of Sine of A/16-series angles are based on following formulae:

It has 4-level of centrals (2^4)

4th central (last central) depends upon 'ab' and number of radicals will be also 4. The form is $90/2 \pm 90/4 \pm 90/8 \pm 90/16$; i.e. $45 \pm 22.5 \pm 11.25 \pm 5.625$. So,

$\sin(45 \pm 22.5 \pm 11.25 \pm 5.625 \pm A/16)$

$= \sin(45 \pm (22.5 \pm (11.25 \pm (5.625 \pm A/16)]$

$= \sqrt{2 \pm \sqrt{2 \pm \sqrt{2 \pm \sqrt{2 \pm 1 \times \sqrt{3}}}}}/2$

3rd central (second-last central) depends upon 'a' and number of radicals will be 3. The form is $45 \pm 22.5 \pm 11.25$. So,

$\sin(45 \pm 22.5 \pm 11.25 \pm A/16)$

$= \sin(45 \pm (22.5 \pm (11.25 \pm A/16]$

$= \sqrt{2 \pm \sqrt{2 \pm \sqrt{2 \pm 1}}}/2$

SERIES METHOD

First and second centrals (earlier all) depends upon 'b' with 3-radicals. The form is 90, 45 and 45 ± 22.5.

First central is 90 for this purpose, $A/16$ is half-series and $90 - A/16$ is complementary to $A/16$. To obtain Sin $A/16$, we need chord of $A/8$ (means, we have 3-radicals). Therefore, Sin $(90 \pm A/16) = \sqrt{(2 \pm \sqrt{(2 + \sqrt{(2 + b)}})}/2$.

Second central is 45 for this purpose. The angle is $45 \pm A/16$, which is an angle in $A/8$-series. Therefore, it will have 3-radicals with dependency b as Sin $(45 \pm A/16) = \sqrt{(2 \pm \sqrt{(2 - \sqrt{(2 + b)}})}/2$.

Therefore, the general pattern for $A/2^n$ series is as follows:

N^{th} *central has 2^n numbers of angles having exact value of Sine depending upon ab with n numbers of radicals in each Sine.*

$$\text{Sin } (45 \pm (22.5 \pm (11.25 \pm \ldots \pm (90/2^n \pm A/2^n)]$$
$$= \sqrt{(2 \pm \sqrt{(2 \pm \sqrt{(2 \pm \ldots \pm \sqrt{(2 \pm ab)}}})}/2$$

$(n-1)^{th}$ central has 2^{n-1} numbers of angles having exact value of Sine depending upon a with n-1 numbers of radicals in each Sine.

$$\text{Sin } (45 \pm (22.5 \pm (11.25 \pm \ldots \pm (90/2^{n-1} \pm A/2^n)]$$
$$= \sqrt{(2 \pm \sqrt{(2 \pm \sqrt{(2 \pm \ldots \pm \sqrt{(2 \pm a)}}})}/2$$

From 1st to $(n-2)^{th}$ central has 2^{n-1} numbers of angles having exact value of Sine depending upon b with n-1 numbers of radicals in each Sine.

$$\text{Sin of series of } A/2^n \text{ is } \sqrt{(2 \pm \sqrt{(2 \pm (\ldots \pm \sqrt{(2 \pm b)}})}/2$$

We have listed 62 formulae till here [2 from A-series, 4 from $A/2$-series, 8 from $A/4$-series, 16 from $A/8$-series and 32 from $A/16$-series. If we take the $A/32$-series, the set of formulae will be 64 making total angles equal to 126. These angles will spread in the difference of maximum of $A/2^{n-1}$ in

EXACT VALUES IN TRIGONOMETRY

the two-wings of center. Taking further 32 angles[9] from A/32-series, the list of Sine of an angle are as follows:

Angle (A)	Sine of Angle A
2.8125−A/32	√(2 −√(2 + √(2 + √(2 + √(2 +ab]/2
5.625−A/16	√(2 −√(2 + √(2 + √(2 +ab]/2
11.25−A/8	√(2 −√(2 + √(2 +ab]/2
A/16	√(2 −√(2 + √(2 +b]/2
2.8125+A/32	√(2 −√(2 + √(2 + √(2 + √(2 −ab]/2
8.4375−A/32	√(2 −√(2 + √(2 + √(2 −√(2 −ab]/2
11.25−A/16	√(2 −√(2 + √(2 +a]/2
22.5−A/4	√(2 −√(2 +ab]/2
A/8	√(2 −√(2 +b]/2
5.625+A/16	√(2 −√(2 + √(2 + √(2 −ab]/2
8.4375+A/32	√(2 −√(2 + √(2 + √(2 −√(2 +ab]/2
14.0625−A/32	√(2 −√(2 + √(2 −√(2 −√(2 +ab]/2
16.875−A/16	√(2 −√(2 + √(2 −√(2 −ab]/2
11.25+A/16	√(2 −√(2 + √(2 −a]/2
45−A/2	√(2 −ab)/2
A/4	√(2 −b)/2
22.5−A/8	√(2 −√(2 +a]/2
14.0625+A/32	√(2 −√(2 + √(2 −√(2 −√(2 −ab]/2
19.6875−A/32	√(2 −√(2 + √(2 −√(2 + √(2 −ab]/2
11.25+A/8	√(2 −√(2 + √(2 −ab]/2
22.5−A/16	√(2 −√(2 + √(2 −b]/2
16.875+A/16	√(2 −√(2 + √(2 −√(2 +ab]/2

---

[9] There are 64 angles in A/32-series. Out of 64, 16 are based on a, 16 on b and 32 are based on ab. Only the angle based on ab have listed here.

SERIES METHOD

$19.6875+A/32$	$\sqrt{(2-\sqrt{(2+\sqrt{(2-\sqrt{(2+\sqrt{(2+ab}]}}}}/2$
$25.3125-A/32$	$\sqrt{(2-\sqrt{(2-\sqrt{(2-\sqrt{(2+\sqrt{(2+ab}]}}}}/2$
$28.125-A/16$	$\sqrt{(2-\sqrt{(2-\sqrt{(2-\sqrt{(2+ab}]}}}/2$
$33.75-A/8$	$\sqrt{(2-\sqrt{(2-\sqrt{(2-ab}]}}/2$
$22.5+A/16$	$\sqrt{(2-\sqrt{(2-\sqrt{(2-b}]}}/2$
$25.3125+A/32$	$\sqrt{(2-\sqrt{(2-\sqrt{(2-\sqrt{(2+\sqrt{(2-ab}]}}}}/2$
$30.9375-A/132$	$\sqrt{(2-\sqrt{(2-\sqrt{(2-\sqrt{(2-\sqrt{(2-ab}]}}}}/2$
$90-A$	$(2-a^2)/2$
$33.75-A/16$	$\sqrt{(2-\sqrt{(2-\sqrt{(2-a}]}}/2$
$22.5+A/8$	$\sqrt{(2-\sqrt{(2-a}]}/2$
$45-A/4$	$\sqrt{(2-a)}/2$
$A/2$	$a/2$
$28.125+A/16$	$\sqrt{(2-\sqrt{(2-\sqrt{(2-\sqrt{(2-ab}]}}}/2$
$30.9375+A/132$	$\sqrt{(2-\sqrt{(2-\sqrt{(2-\sqrt{(2-\sqrt{(2+ab}]}}}}/2$
$36.5625-+A/32$	$\sqrt{(2-\sqrt{(2-\sqrt{(2+\sqrt{(2-\sqrt{(2+ab}]}}}}/2$
$39.375-A/16$	$\sqrt{(2-\sqrt{(2-\sqrt{(2+\sqrt{(2-ab}]}}}/2$
$33.75+A/16$	$\sqrt{(2-\sqrt{(2-\sqrt{(2+a}]}}/2$
$22.5+A/4$	$\sqrt{(2-\sqrt{(2-ab}]}/2$
$45-A/8$	$\sqrt{(2-\sqrt{(2-b}]}/2$
$36.5625-+A/32$	$\sqrt{(2-\sqrt{(2-\sqrt{(2+\sqrt{(2-\sqrt{(2-ab}]}}}}/2$
$42.1875-A/32$	$\sqrt{(2-\sqrt{(2-\sqrt{(2+\sqrt{(2+\sqrt{(2-ab}]}}}}/2$
$33.75+A/8$	$\sqrt{(2-\sqrt{(2-\sqrt{(2+ab}]}}/2$
$45-A/16$	$\sqrt{(2-\sqrt{(2-\sqrt{(2+b}]}}/2$
$39.375+A/16$	$\sqrt{(2-\sqrt{(2-\sqrt{(2+\sqrt{(2+ab}]}}}/2$
$42.1875+A/32$	$\sqrt{(2-\sqrt{(2-\sqrt{(2+\sqrt{(2+\sqrt{(2+ab}]}}}}/2$
$47.8125-A/32$	$\sqrt{(2+\sqrt{(2-\sqrt{(2+\sqrt{(2+\sqrt{(2+ab}]}}}}/2$
$50.625-A/16$	$\sqrt{(2+\sqrt{(2-\sqrt{(2+\sqrt{(2+ab}]}}}/2$

EXACT VALUES IN TRIGONOMETRY

56.25-A/8	$\sqrt{(2+\sqrt{(2-\sqrt{(2+ab]}}}/2$
45+A/16	$\sqrt{(2+\sqrt{(2-\sqrt{(2+b]}}}/2$
47.8125+A/32	$\sqrt{(2+\sqrt{(2-\sqrt{(2+\sqrt{(2+\sqrt{(2-ab]}}}}}/2$
53.3125-A/32	$\sqrt{(2+\sqrt{(2-\sqrt{(2+\sqrt{(2-\sqrt{(2-ab]}}}}}/2$
56.25-A/16	$\sqrt{(2+\sqrt{(2-\sqrt{(2+a]}}}/2$
67.5-A/4	$\sqrt{(2+\sqrt{(2-ab]}}/2$
45+A/8	$\sqrt{(2+\sqrt{(2-b]}}/2$
50.625+A/16	$\sqrt{(2+\sqrt{(2-\sqrt{(2+\sqrt{(2-ab]}}}}/2$
53.3125+A/32	$\sqrt{(2+\sqrt{(2-\sqrt{(2+\sqrt{(2-\sqrt{(2+ab]}}}}}/2$
59.0625-A/132	$\sqrt{(2+\sqrt{(2-\sqrt{(2-\sqrt{(2-\sqrt{(2+ab]}}}}}/2$
61.875-A/16	$\sqrt{(2+\sqrt{(2-\sqrt{(2-\sqrt{(2-ab]}}}}/2$
56.25+A/16	$\sqrt{(2+\sqrt{(2-\sqrt{(2-a]}}}/2$
67.5-A/8	$\sqrt{(2+\sqrt{(2-a]}}/2$
45+A/4	$\sqrt{(2+a)}/2$
A	ab/2
90-A/2	b/2
59.0625+A/132	$\sqrt{(2+\sqrt{(2-\sqrt{(2-\sqrt{(2-\sqrt{(2-ab]}}}}}/2$
64.6875-A/32	$\sqrt{(2+\sqrt{(2-\sqrt{(2-\sqrt{(2+\sqrt{(2-ab]}}}}}/2$
56.25+A/8	$\sqrt{(2+\sqrt{(2-\sqrt{(2-ab]}}}/2$
67.5-A/16	$\sqrt{(2+\sqrt{(2-\sqrt{(2-b]}}}/2$
61.875+A/16	$\sqrt{(2+\sqrt{(2-\sqrt{(2-\sqrt{(2+ab]}}}}/2$
64.6875+A/32	$\sqrt{(2+\sqrt{(2-\sqrt{(2-\sqrt{(2+\sqrt{(2+ab]}}}}}/2$
70.3125-A/32	$\sqrt{(2+\sqrt{(2+\sqrt{(2-\sqrt{(2+\sqrt{(2+ab]}}}}}/2$
73.125-A/16	$\sqrt{(2+\sqrt{(2+\sqrt{(2-\sqrt{(2+ab]}}}}/2$
78.75-A/8	$\sqrt{(2+\sqrt{(2+\sqrt{(2-ab]}}}/2$
67.5+A/16	$\sqrt{(2+\sqrt{(2+\sqrt{(2-b]}}}/2$
70.3125+A/32	$\sqrt{(2+\sqrt{(2+\sqrt{(2-\sqrt{(2+\sqrt{(2-ab]}}}}}/2$

SERIES METHOD

$75.9375 - A/32$	$\sqrt{(2 + \sqrt{(2 + \sqrt{(2 - \sqrt{(2 - \sqrt{(2 - ab}}}}]/2$
$78.75 - A/16$	$\sqrt{(2 + \sqrt{(2 + \sqrt{(2 - a}}}]/2$
$45 + A/2$	$\sqrt{(2 + ab)}/2$
$90 - A/4$	$\sqrt{(2 + b)}/2$
$67.5 + A/8$	$\sqrt{(2 + \sqrt{(2 + a}}]/2$
$73.125 + A/16$	$\sqrt{(2 + \sqrt{(2 + \sqrt{(2 - \sqrt{(2 - ab}}}}]/2$
$75.9375 + A/32$	$\sqrt{(2 + \sqrt{(2 + \sqrt{(2 - \sqrt{(2 - \sqrt{(2 + ab}}}}}]/2$
$81.5625 - A/32$	$\sqrt{(2 + \sqrt{(2 + \sqrt{(2 + \sqrt{(2 - \sqrt{(2 + ab}}}}}]/2$
$84.375 - A/16$	$\sqrt{(2 + \sqrt{(2 + \sqrt{(2 + \sqrt{(2 - ab}}}}]/2$
$78.75 + A/16$	$\sqrt{(2 + \sqrt{(2 + \sqrt{(2 + a}}}]/2$
$67.5 + A/4$	$\sqrt{(2 + \sqrt{(2 + ab}}]/2$
$90 - A/8$	$\sqrt{(2 + \sqrt{(2 + b}}]/2$
$81.5625 + A/32$	$\sqrt{(2 + \sqrt{(2 + \sqrt{(2 + \sqrt{(2 - \sqrt{(2 - ab}}}}}]/2$
$87.1875 - A/32$	$\sqrt{(2 + \sqrt{(2 + \sqrt{(2 + \sqrt{(2 + \sqrt{(2 - ab}}}}}]/2$
$78.75 + A/8$	$\sqrt{(2 + \sqrt{(2 + \sqrt{(2 + ab}}}]/2$
$90 - A/16$	$\sqrt{(2 + \sqrt{(2 + \sqrt{(2 + b}}}]/2$
$84.375 + A/16$	$\sqrt{(2 + \sqrt{(2 + \sqrt{(2 + \sqrt{(2 + ab}}}}]/2$
$87.1875 + A/32$	$\sqrt{(2 + \sqrt{(2 + \sqrt{(2 + \sqrt{(2 + \sqrt{(2 + ab}}}}}]/2$

From above, we conclude, exact value of Sine of an angle is based on chord (a), supplementary chord (b) and formula operator $\sqrt{(2}$. From these two chords, we can calculate Sine of any angles within the $A/2^n$-series for academic, for practical or for programming purpose in the given route of angle having exactly known chord.

In each step, we see, the result has divided by 2. If the result before division by 2, is the chord of double angle. Therefore, using above chord, we can calculate any trigonometric ratios for that angle within the series. Hence, we calculate following numbers of *trigonometric ratios of an angle*:

45

EXACT VALUES IN TRIGONOMETRY

- For A- series- A 2^1 2   2
- For A/2- series 2^2 4   6
- For A/4- series 2^3 8   14
- For A/8- series 2^4 16   30
- For A/16- series 2^5 32   62
- For A/32- series 2^6 64   126
- ....
- For A/2^n- series 2^{n+1} $2^{n+1} - 2$

Sine of any angle

In the above process, we obtained exact value of Sine (and other trigonometric ratios) of angles within the series. May we able to find such exact value of any desired angle? Theoretical answer is 'no'. However, practically, the answer is 'yes' for an *appropriate accuracy*. For example, for 1°, this does not fall in series. In the contrast, 0.99999999....975 falls in A/2^n-series. If you want 1000 or 10^9-digits of accuracy for Sin 1°, you shall obtain exact value for Sin 0.99999999...1000-digits or 10^9-digits ....975°. In this case, *appropriate accuracy* is 1000-digits or 10^9 digits or any of user-defined limit.

Other trigonometric relations than Sine of an angle

In above process, we obtained chord (a) for an angle. If we change the sign of first radical, the result is supplementary chord (b). For example, 'a' for double of 75.9375+A/32 is $\sqrt{(2 + \sqrt{(2 + \sqrt{(2 - \sqrt{(2 - \sqrt{(2 + ab]}}}}}$ and $\sqrt{(2 - \sqrt{(2 + \sqrt{(2 - \sqrt{(2 - \sqrt{(2 + ab]}}}}}$ will be 'b'.

Once, we have chords (a, b, and ab), we can calculate all trigonometric ratios (Cos A, Tan A etc.) using formulae for each trigonometric relation given on page 2.

Chapter 4
ANGLE-REWRITTEN METHOD

KEY WORDS – Central, half-series, Appropriate accuracy, Series Sum-Difference, Central-count, Averaging, Binary-Sign.

4.1 Exact Sine of an Angle

Exact value of Sine of an angle under *Angle-Rewritten method* partially depends upon the concept of *Series-Method*. In the Series-method, we calculate Sine of angles within a series, but in *Angle-Rewritten method*, individual angle is base.

In general, if any angle is perfectly divisible by $A/2^n$ (where A is an angle of known chord), then exact value of Sine of that angle is possible. The smallest angle in the series will be $A/2^{n+1}$, and multiple of $180/A^{th}$ item of the $A/2^n$-series will be depends upon 90° only.

47

EXACT VALUES IN TRIGONOMETRY

For example, in 60°/32-series, smallest angle whose exact trigonometric relations are available will be 60°/64= 0.9375°. All the angles which are multiple of 0.9375° have exact value. In the general terms, all multiples of $A/2^n$ has exact value.

Multiples of 180/60 = 3 in the series will not depend upon chord of 60°. So, 0.9375°×3= 2.8125° or 0.9375°×30= 28.125° is not depend upon 1, √3, rather depends upon √2. In case of 36°, 180/36=5; any multiples of 5 is not depends upon 'a' or 'b' of 36°, rather depends upon √2 itself.

Further example for $11.2675781249998°/(36/2^{12})$ =1282, which is an integer and we know the chord of 36°. We can find the exact value of 11.2675781249998°. Another example, product of 29127 with $(36/2^{20})$ is 0.999996185 and 29128 with $(36/2^{20})$ is 1.000030518. We can determine the exact value of either Sin 0.999996185° or Sin 1.000030518°.

How to write an angle under Angle-Rewritten method

In Angle-Rewritten Method, the angle is written such that each *central* (90°-route angle) is either subtracted (to obtain lower angle) or added (to obtain higher angle) the half of earlier *central* initiating from 45° with bracket in each step. The first *central* is 45, the second is 22.5, and the third is 11.25 and so on. In this method, 22.5° is written as 45°-(22.5°). For 56.25°, it is written as 45°+(22.5°-(11.25°][10].

In addition, if the angle is in any series of $A/2^n$, Angle-Rewritten method has used taking ± $A/2^n$ in the last item of above series without bracket. For example, 23.4375 = 22.5+0.9375= 45-(22.5-0.9375) or 55.3125 = 56.25-0.9375 = 45°+(22.5°-(11.25°-0.9375].

---

[10] To make the presentation easy, all the closing brackets are collapsed by single bracket ']' in this Chapter.

ANGLE-REWRITTEN METHOD

One major point is that, the last power-base in central and $A/2^n$ requires to be same. If they are different, we need to adjust $A/2^n$ for same base taking half angle. Otherwise, we should reconsider different dependency of chord (a) or supplementary chord (b) or the product of ab. For example, 23.4375 = 22.5+0.9375= 45-(22.5-0.9375) = 90/2-(90/4-60/64). The last central is based on 90/4 whereas angle is 60/64. The power-base is different. To make it simple, we may convert 60/64 to such an angle which has power-base of 4; that could be 60/64 = 30/32 = 15/16 = 7.5/8 = 3.75/8 = 1.875/4.

So, 23.4375 = 22.5 +0.9375 = 45-(22.5-0.9375) = 90/2-(90/4 - 1.875/4).

4.1.1 Half-Series

To calculate the last items of the Angle-Rewritten method, we need to have knowledge of half-series. Many cases, the given angle (A) itself may within half-series. For this, first of all, we need to find whether the given angle is in the series of half (i.e. $A/2, A/4, A/8,...$) of angle having known chord. For example, 15° is 60/4, so, is a member of half-series.

Angle-Rewritten method: Half-series technique

In Angle-Rewritten method for half-series, the angle is written based on half of earlier number along with a bracket. Each number will be replaced by a radical of 2 except last one, which is replaced by supplementary chord (b). in case, we need chord of the given angle, we need to apply Angle-Rewritten for the given angle (A). if we need Sin A, we need to apply Angle-Rewritten for double of given angle (2A), because Sin A is half of chord of 2A. *For example:*

11.25= 45-(22.5+(11.25]   (90°-route: $b=\sqrt{2}$)

EXACT VALUES IN TRIGONOMETRY

Crd $11.25° = \sqrt{(2 - \sqrt{(2 + \sqrt{2}})}/2$

For Sin 11.25°, we need to compute Angle-Rewritten for **22.5°** as 45-(22.5). So, Sin $11.25° = \sqrt{(2 - \sqrt{2})}/2$

For Sine of 7.5, we need to apply Angle-Rewritten for 15° = 60-(30+15]. So, Sin $7.5° = \sqrt{(2 - \sqrt{(2 + \sqrt{3}})}/2$.

Half-series is major contributor for exact value of Sine either under Series-method of under Angle-Rewritten method. For detail, we need half-series of all the angles divisible by 3. Few most common angles are given hereunder. Main purpose of these series is to expose the pattern of formulae. Need to say, half of 'a' is Sine of half-angle in each case. Half of 'b' is Sine of complementary angle of that half-angle in each case.

The most useful is route-90:

2^n	A	'a'	'b'
0	90	$\sqrt{2}$	$\sqrt{2}$
1	45	$\sqrt{(2-\sqrt{2})}$	$\sqrt{(2+\sqrt{2})}$
2	22.5	$\sqrt{(2-\sqrt{(2+\sqrt{2}})}$	$\sqrt{(2+\sqrt{(2+\sqrt{2}})}$
3	11.25	$\sqrt{(2-\sqrt{(2+\sqrt{(2+\sqrt{2}})})}$	$\sqrt{(2+\sqrt{(2+\sqrt{(2+\sqrt{2}})})}$
4	5.625	$\sqrt{(2-\sqrt{(2+\sqrt{(2+\sqrt{(2+\sqrt{2}})})})}$	$\sqrt{(2+\sqrt{(2+\sqrt{(2+\sqrt{(2+\sqrt{2}})})})}$
5	2.8125	$\sqrt{(2-\sqrt{(2+\sqrt{(2+\sqrt{(2+\sqrt{(2+\sqrt{2}})})})})}$	$\sqrt{(2+\sqrt{(2+\sqrt{(2+\sqrt{(2+\sqrt{(2+\sqrt{2}})})})})}$
6	1.40625	$\sqrt{(2-\sqrt{(2+\sqrt{(2+\sqrt{(2+\sqrt{(2+\sqrt{(2+\sqrt{2}})})})})})}$	$\sqrt{(2+\sqrt{(2+\sqrt{(2+\sqrt{(2+\sqrt{(2+\sqrt{(2+\sqrt{2}})})})})})}$
7	0.703125	$\sqrt{(2-\sqrt{(2+\sqrt{(2+\sqrt{(2+\sqrt{(2+\sqrt{(2+\sqrt{(2+\sqrt{2}})})})})})})}$	$\sqrt{(2+\sqrt{(2+\sqrt{(2+\sqrt{(2+\sqrt{(2+\sqrt{(2+\sqrt{(2+\sqrt{2}})})})})})})}$

Route-60

2^n	A	'a'	'b'
0	60	1	$\sqrt{3}$
1	30	$\sqrt{(2-\sqrt{3})}$	$\sqrt{(2+\sqrt{3})}$
2	15	$\sqrt{(2-\sqrt{(2+\sqrt{3}})}$	$\sqrt{(2+\sqrt{(2+\sqrt{3}})}$
3	7.5	$\sqrt{(2-\sqrt{(2+\sqrt{(2+\sqrt{3}})})}$	$\sqrt{(2+\sqrt{(2+\sqrt{(2+\sqrt{3}})})}$
4	3.75	$\sqrt{(2-\sqrt{(2+\sqrt{(2+\sqrt{(2+\sqrt{3}})})})}$	$\sqrt{(2+\sqrt{(2+\sqrt{(2+\sqrt{(2+\sqrt{3}})})})}$

ANGLE-REWRITTEN METHOD

n	A	'a'	'b'
5	1.875	$\sqrt{(2-\sqrt{(2+\sqrt{(2+\sqrt{(2+\sqrt{3}]}}}}$	$\sqrt{(2+\sqrt{(2+\sqrt{(2+\sqrt{(2+\sqrt{3}]}}}}$
6	0.9375	$\sqrt{(2-\sqrt{(2+\sqrt{(2+\sqrt{(2+\sqrt{(2+\sqrt{3}]}}}}}$	$\sqrt{(2+\sqrt{(2+\sqrt{(2+\sqrt{(2+\sqrt{(2+\sqrt{3}]}}}}}$
7	0.46875	$\sqrt{(2-\sqrt{(2+\sqrt{(2+\sqrt{(2+\sqrt{(2+\sqrt{(2+\sqrt{3}]}}}}}}$	$\sqrt{(2+\sqrt{(2+\sqrt{(2+\sqrt{(2+\sqrt{(2+\sqrt{(2+\sqrt{3}]}}}}}}$

Route-72

n	A	'a'	'b'
0	72	$(\sqrt{(10-\sqrt{20})}/2)$	$(\sqrt{5}+1)/2$
1	36	$\sqrt{(2-(\sqrt{5}+1)/2)}$	$\sqrt{(2+(\sqrt{5}+1)/2)}$
2	18	$\sqrt{(2-\sqrt{(2+(\sqrt{5}+1)/2]}}$	$\sqrt{(2+\sqrt{(2+(\sqrt{5}+1)/2]}}$
3	9	$\sqrt{(2-\sqrt{(2+\sqrt{(2+(\sqrt{5}+1)/2]}}}$	$\sqrt{(2+\sqrt{(2+\sqrt{(2+(\sqrt{5}+1)/2]}}}$
4	4.5	$\sqrt{(2-\sqrt{(2+\sqrt{(2+\sqrt{(2+(\sqrt{5}+1)/2]}}}}$	$\sqrt{(2+\sqrt{(2+\sqrt{(2+\sqrt{(2+(\sqrt{5}+1)/2]}}}}$
5	2.25	$\sqrt{(2-\sqrt{(2+\sqrt{(2+\sqrt{(2+\sqrt{(2+(\sqrt{5}+1)/2]}}}}}$	$\sqrt{(2+\sqrt{(2+\sqrt{(2+\sqrt{(2+\sqrt{(2+(\sqrt{5}+1)/2]}}}}}$
6	1.125	$\sqrt{(2-\sqrt{(2+\sqrt{(2+\sqrt{(2+\sqrt{(2+\sqrt{(2+(\sqrt{5}+1)/2]}}}}}}$	$\sqrt{(2+\sqrt{(2+\sqrt{(2+\sqrt{(2+\sqrt{(2+(\sqrt{5}+1)/2]}}}}}}$
7	0.5625	$\sqrt{(2-\sqrt{(2+\sqrt{(2+\sqrt{(2+\sqrt{(2+\sqrt{(2+\sqrt{(2+(\sqrt{5}+1)/2]}}}}}}}$	$\sqrt{(2+\sqrt{(2+\sqrt{(2+\sqrt{(2\sqrt{(2++\sqrt{(2+\sqrt{(2+(\sqrt{5}+1)/2]}}}}}}}$

Route-12

n	A	'a'	'b'
0	12	$(\sqrt{(30-\sqrt{180})}-\sqrt{5}-1)/4$	$\sqrt{(28+\sqrt{180}-\sqrt{20}+2\sqrt{(30-\sqrt{180})}(\sqrt{5}+1))}/4$
1	6	$\sqrt{(2-\sqrt{(28+\sqrt{180}-\sqrt{20}+2\sqrt{(30-\sqrt{180})}(\sqrt{5}+1))/4)}}$	$\sqrt{(2+\sqrt{(28+\sqrt{180}-\sqrt{20}+2\sqrt{(30-\sqrt{180})}(\sqrt{5}+1))/4)}}$
2	3	$\sqrt{(2-\sqrt{(2+\sqrt{(28+\sqrt{180}-\sqrt{20}+2\sqrt{(30-\sqrt{180})}(\sqrt{5}+1))/4]}}}$	$\sqrt{(2+\sqrt{(2+\sqrt{(28+\sqrt{180}-\sqrt{20}+2\sqrt{(30-\sqrt{180})}(\sqrt{5}+1))/4]}}}$
3	1.5	$\sqrt{(2-\sqrt{(2+\sqrt{(2+\sqrt{(28+\sqrt{180}-\sqrt{20}+2\sqrt{(30-\sqrt{180})}(\sqrt{5}+1))/4]}}}}$	$\sqrt{(2+\sqrt{(2+\sqrt{(2+\sqrt{(28+\sqrt{180}-\sqrt{20}+2\sqrt{(30-\sqrt{180})}(\sqrt{5}+1))/4]}}}}$
4	0.75	$\sqrt{(2-\sqrt{(2+\sqrt{(2+\sqrt{(2+\sqrt{(28+\sqrt{180}-\sqrt{20}+2\sqrt{(30-\sqrt{180})}(\sqrt{5}+1))/4]}}}}}$	$\sqrt{(2+\sqrt{(2+\sqrt{(2+\sqrt{(2+\sqrt{(28+\sqrt{180}-\sqrt{20}+2\sqrt{(30-\sqrt{180})}(\sqrt{5}+1))/4]}}}}}$
5	0.375	$\sqrt{(2-\sqrt{(2+\sqrt{(2+\sqrt{(2+\sqrt{(2+\sqrt{(28+\sqrt{180}-\sqrt{20}+2\sqrt{(30-\sqrt{180})}(\sqrt{5}+1))/4]}}}}}}$	$\sqrt{(2+\sqrt{(2+\sqrt{(2+\sqrt{(2+\sqrt{(28+\sqrt{180}-\sqrt{20}+2\sqrt{(30-\sqrt{180})}(\sqrt{5}+1))/4]}}}}}$

4.1.2 Varieties of Techniques for Angle-Rewritten

Half-series gives just half of an angle. In case, these halves of different routes were added or subtracted, then the chord or Sine of that angle become a bit complex. We need to bifurcate these routes to determine chord or sine of that angle. There would be variety of techniques to determine chord or Sine of that angle. They may be (but not limited to):

I.   Series Sum-Difference technique

In this technique, first of all, route-angle of given angle is ascertained. Then a sum and difference table for 90 and that route-angle is prepared for each halves. Let's explain Series Sum-Difference technique through Sin 37.5^0 and Sin 7.5°. Both of these angles falls in route-60; because 60/37.5=1.6 and 60/7.5 = 8, both are exact value for 2^n, ignoring the decimal point.

			Sum	Difference
A	90.00	60.00	150.00	30.00
A/2	45.00	30.00	75.00	15.00
A/4	22.50	15.00	**37.50**	**7.50**
A/8	11.25	7.50	18.75	3.75

Our intended angles 37.5^0 (sum) and Sin 7.5° (difference) falls in A/4-series. The representing format will be 22.5°+A/4 and 22.5°-A/4 (all have power-base of 4, so dependency is 'ab'). *If we see the half-series, 7.5° is in A/8 for 60°. Its exact Sine is possible under half-series technique too. For the comparative study, this angle has repeated here.*

Here, 37.5°= 45 – (22.5 – 15) = 90/2 – (90/4 – 60/4)

Hence, Sin 37.5° = √(2 -√(2 -ab]/2 =√(2 -√(2 -1×√3]/2 = √(2 -√(2 -√3]/2

Next, 7.5° = 45 – (22.5 + 15) = 90/2 – (90/4 + 60/4) Hence, Sin 7.5° = √(2 -√(2 +ab]/2 =√(2 -√(2 +1×√3]/2 = √(2 -√(2 + √3]/2.

Let us take another example of Sin 55.3125°. For this, firstly, find the route-angle. Doubling 55.3125°, its angle-route would be either 75° or 60° or 30° as:

Double angle	Reducing to first quadrant
55.3125	55.3125
110.625	20.625
221.25	41.25
442.5	82.5
885	75
1770	60
3540	30

For this, firstly checking 60/32 and subtract from given angle. If the remaining angle is divisible by 90/32. Since both route-angles are in same power-base of 32, this is our target angle. Rewriting the given angle as (care the writing pattern under Angle-Rewritten method, always bracket before any number except A/32, and always in a half-order):

=45+(22.5-(11.25+(5.625-(**2.8125+60/32**]

Here, both 90 and 60 ends on same power-base of 2^5, so dependency is 'ab'. The formula for Sin =√[2 + √(2 -√(2 + √(2 -√(2 + ab]]/2

=√[2 + √(2 -√(2 + √(2 -√(2 +√3]]/2

Let us take an example of 0.9375°; Checking for divisibility, 60/0.9375 is 64. So, this is a member of route-60.

EXACT VALUES IN TRIGONOMETRY

Firstly, find the *central* of half-series, which correlates the given angle[11]. For example, if we need value of Sin 0.9375°, *central* near to this would be either 90/32 (2.8125) or 90/16 (5.625) or something else.

Secondly, determine sum-difference (±A) in these *central* taking same divisor (e.g. with 60°). In our case of 2.8125 is based on divisor 32 and 5.625 is based on divisor 16. So,

$$2.8125 \pm 60/32 \quad = 4.6875 \text{ and } 0.9375$$
$$5.625 \pm 60/16 \quad = 9.3750 \text{ and } 1.8750$$

We obtained 0.9375 is 2.8125 - 60/32; we shall take this value to determine Sin 0.9375° (even there are other ways also). Since, both part of formula has same 2^n-base, our formula operator is 'ab'. Writing the angle in Angle-Rewritten method as 45-(22.5+(11.25+(5.625+(2.8125+60/32)

So, $\text{Sin } 0.9375° = \sqrt{[2-\sqrt{(2 + \sqrt{(2 + \sqrt{(2 + \sqrt{(2 + ab)}}}}}]/2$
$$= \sqrt{[2-\sqrt{(2 + \sqrt{(2 + \sqrt{(2 + \sqrt{(2 + \sqrt{3})}}}}}]/2$$

II. Averaging technique

In the sum-difference technique, the angle and centrals, were in pattern of same power-base (2^n). We can calculate same angle more differently or above calculation may be interpreted differently as *averaging for Angle-Rewritten method*. This technique is the best way to find the exact value of un-patterned angles. Taking example of Sin 76.875°:

- Divide the angle by $60/2^n$ and find the lowest integer quotient (resulting quotient is 41). Since the angle is perfectly divisible by $60/2^n$, there is exact value of Sine

---

[11] Finding central is simple, we know 180/60= 3 or 180/36= 5. So, angle which is multiples of 3 of $60/2^n$ or multiple of 5 of $36/2^n$ is central for this purpose.

ANGLE-REWRITTEN METHOD

in 60°-route. The lowest value is 60/32 in this series and the series itself is 60/16.

76.875	64	82
76.875	32	**41**
76.875	16	20.5

- In 60°-route, multiple of 3 is not dependent upon chord of 60°. Find the nearest 3-divisible number near to the lowest quotient (42 is the nearest number divisible by 3). We know our intended angle is 42-1=41, which is based on 60/32. Our series is based on 60/16, to make the angle in 60/16-series, 39×60/16+60/16 is the required central.

If the quotient under 60°-route were divisible by 3, e.g. 42, the Sine of such angle will depend upon √2 only;

If the quotient is ±1 from multiples of 3, take the central just higher or lower than calculated central to make it ±2; e.g., our calculated central is 42 times and intended angle was 41 times. To determine sine of intended angle 39+2 is the desired angle. So, central is 39 and +2 is the angle.

- Write the angle for the Angle-Rewritten method based on average of 0 and 90 to obtain target angle as:

$$76.875$$
$$\underline{3.75} \quad \text{-ve}$$
$$73.125$$

		Average	Rewritten
0	90	45	45
90	45	67.5	22.5
90	67.5	78.75	11.25
67.5	78.75	73.125	-5.625

So, 76.875=45+(22.5+(11.25-(5.625-3.75])

Sin 76.875=√(2 + √(2 + √(2 -√(2 -√(3]/2

III. Central-count technique

All the centrals (i.e. route-90 angles), as we see above, depends upon the $\sqrt{2}$. We can replace each central by this value adjusting route-angle in the last. Let's take example of Sin 4.6875° with a bit different presentation.

Firstly, finding the root of angle as $(90+60)/4.6875 = 32$. This means there are five 2's in route-60°.

$4.6875 = (45-(22.5+(11.25+(5.625\underline{-60/16}]/2.$

Converting each part under 90° to $\sqrt{2}$ and 60° to $1 \times \sqrt{3}$, Sin $4.6875° = \sqrt{(2 - \sqrt{(2 + \sqrt{(2 + \sqrt{(2 + \sqrt{(2 - \underline{1 \times \sqrt{3}})}}}}}]/2$.

What's about complementary? The technique is exactly similar. The angle will be broken up as:

$90 - 4.6875 = 85.3125$

$= \underline{90-(45-}(22.5+(11.25+(5.625-60/16]/2.$

$= \underline{(180-45+}(22.5+(11.25+(5.625-60/16]/2.$

$= \underline{(135+}(22.5+(11.25+(5.625-60/16]/2.$

$= \underline{(45+}(22.5+(11.25+(5.625\underline{-60/16}]/2.$

So, Sin 85.3125° $= = \sqrt{(2 + \sqrt{(2 + \sqrt{(2 + \sqrt{(2 + \sqrt{(2 - \underline{1 \times \sqrt{3}}}}}}}/2$.

IV. Binary-Sign technique

Angle-Rewritten method is simple for understanding and use. However, its calculation of angles seems cumbersome, especially on sign within multiple brackets. There is simple way to remove these difficulties in Binary-Sign technique.

Let us take an example of 8.4375°. Subtracting 5.625 from 8.4375, we obtain 2.8125 (a route-90 angle). Take the centrals from 90 by halving it to obtain 8.4375. Find the quotient dividing first central to second central. The result is always ±2. Our target is not to the result, but to the sign only (so, named as binary-sign).

ANGLE-REWRITTEN METHOD

		Binary-sign
90		
45	45	2
-22.5	22.5	-2
-11.25	11.25	2
-5.625	5.625	2
2.8125	8.4375	-2

Just taking the sing from 'Binary-sign' column initiating from the top as +2 is +√(2 and – 2 is – √(2:

Sin 8.4375°= √(2 – √(2 +√(2 +√(2 – √2]/2.

Binary-Sign technique resolves all the number and sign complexity.

4.2 General procedures - exact value

We can perform certain test to determine exact value of Sine or its desired appropriate accuracy.

- Only if the sum of digits of given angle (A) is divisible by 3 [Perfectly divisible by 3] may have exact value of Sine. Furthermore, if the last 2-digits after decimal is divisible by 25, the angle has exact value of Sine.
- Firstly, find the route-angle- 90°, 60° or 36°. This can be extended to 90±60 or 90±36 or complementary of any of above. Furthermore, these can be extended to halves and doubles including complementary. Therefore, we may have route-angles 90°, 60°, 36°, 30°, 54°, 150°, 126°, 72°, 18°, 9°, 81°, 15°, 75° etc.
- Secondly, test whether the angle is in half-series. If it falls in the half-series, apply Angle-Rewritten method.
- Thirdly, if the angle is not in half-series but have exact value of Sine, test the angle as:
  - Subtract the nearest central from the given angle (A). For example, if the angle is 8.4375, subtract 5.625. The result will be 2.8125. In other way, we can subtract the

EXACT VALUES IN TRIGONOMETRY

angle (A) from 11.25, again the result if 2.8125. If the angle (A) is 4.6875, subtract 2.8125, the result will be 0.9375 or alternatively the given angle (A) can be subtracted from 5.625 too. If the angle (A) is 11.8125, subtract 11.25, the result is 0.5625.

- Carefully examine the result whether the angle is in half-series of any angle of 90, 60 or 36. In our cases, 2.8125, 0.9375 and 0.5625 are in half-series of 90, 60 and 36 respectively. In case, the result is not in half-series but further central can be subtracted, subtract the second central too. For example, for 15.9375, nearest central is 11.25. Subtracting the result is 4.6875. This is not in half-series, but we can subtract new central of 2.8125 resulting reminder is 1.875, a half-series member of route-60.
- Compare the power-base of 2^n in central and remaining angle. In case, the power-base of 2^n is same for central and given angle both, use the Angle-Rewritten method for the centrals for Sin A.
- In case power-base of 2^n is different in central and given angle, change the given angle (A) to same base of n in central using half-series[12]. Pick up the chords (a and b) of that new base. If the chord is not availed, calculate using addition, subtraction, or half-series.
- From these Angle-rewritten angles, we replace the angle by $\sqrt{(2}$ in each stage of angle except last case, where no bracket is used. For this, binary-sign technique is recommended to reduce the confusion of sign and brackets.

---

[12] To determine the route-angle, double the given angle in series. In case the product is more than 90, double the reminder only. The integer angle will be the route-angle. As already explained, same angle may have more than one route-angle.

- Fourthly, if the given angle (A) is not divisible by 3 (means the angle does not fall within any route-angle as above), take the nearest 3-divisible angle (say A') *with desired appropriate accuracy*[13]. Use the Angle-Rewritten method for A'. To determine Sin A, add A-A' as √(2 ± √(2 ± .... ± √(2)) .. ± (A - A'). In this case, radian measurement requires for A-A' (not for other angles). In this case, recommended method is *Precise-Rewritten Method* (see Chapter 5) than as described here. For example, 40° is not perfectly divisible by 3. Recommended procedure is *Precise-Rewritten Method*. Explaining above examples:

Ex-i Sin 0.0439453125°. Testing the half-series in route-90, $90/2^{11}$= 90/2048 =0.0439453125; so, it is in half-series of route -90 with and the operator is √2. It shall contain 11 radicals for 2.

Applying Angle-Rewritten method (Averaging- technique)

		Average	Increment
0	90	45	45
0	45	22.5	-22.5
0	22.5	11.25	-11.25
0	11.25	5.625	-5.625
0	5.625	2.8125	-2.8125
0	2.8125	1.40625	-1.40625
0	1.4062500000	0.7031250000	-0.70313
0	0.7031250000	0.3515625000	-0.35156
0	0.3515625000	0.1757812500	-0.17578
0	0.1757812500	0.0878906250	-0.08789
0	0.0878906250	0.0439453125	-0.04395

---

[13] Say someone took 20° as the angle A and central as 18°, Sin A will not be acceptable accurate; or similar the case of 19.8° or 19.92° too. For 20°, an acceptable accuracy would be not less than 19.9999 (at least 4 decimal places of accuracy). So desired appropriate accuracy is one of the critical factor for actual accuracy.

EXACT VALUES IN TRIGONOMETRY

Hence, $0.0439453125 = 45 - (22.5 + (11.25 + 5.625 + (2.8125 + (1.40625 + (0.703125 + (0.3515625 + (0.17578125 + (0.087890625 + 0.0439453125]$.

$\text{Sin } 0.0439453125° = \sqrt{(2 - \sqrt{(2 + \sqrt{(2 + \sqrt{(2 + \sqrt{(2 + \sqrt{(2 + \sqrt{(2 + \sqrt{(2 + \sqrt{(2 + \sqrt{(2)}}}}}}}}}]/2$.

For half-series, we may just write the formula. There shall be n number of 2's with radicals for route-90. So, we can write the Sine of 0.0439453125° as $\sqrt{(2 - \sqrt{(2 + \sqrt{(2 + \sqrt{(2 + \sqrt{(2 + \sqrt{(2 + \sqrt{(2 + \sqrt{(2 + \sqrt{(2 + \sqrt{2}}}}}}}}}]/2$ directly.

Alternative way as Binary-sign technique

Binary-Sign technique is easy to handle the sign as:

Given angle	0.043945313	
90	90	Sign
45	45	2
-22.5	22.5	-2
-11.25	11.25	2
-5.625	5.625	2
-2.8125	2.8125	2
-1.40625	1.40625	2
-0.703125	0.703125	2
-0.3515625	0.3515625	2
-0.17578125	0.17578125	2
-0.087890625	0.087890625	2
-0.0439453125	0.0439453125	2

Taking the multiples of 2 with sign, we can write Sin $0.0439453125° = \sqrt{(2 - \sqrt{(2 + \sqrt{(2 + \sqrt{(2 + \sqrt{(2 + \sqrt{(2 + \sqrt{(2 + \sqrt{(2 + \sqrt{(2 + \sqrt{(2 + \sqrt{2}}}}}}}}}}]/2$

Ex-ii Sin 0.9375°. Testing the half-series in route-60, $60/2^6 = 0.0.9375$; so, it is in half-series of route -60.

We can directly use half-series under series technique or other formula given in this Chapter with 6 no. of 2's as:

ANGLE-REWRITTEN METHOD

Sin 0.9375° = $\sqrt{(2 - \sqrt{(2 + \sqrt{(2 + \sqrt{(2 + \sqrt{(2 + \sqrt{3})})})})}]/2}$

In this case, $A/2^n$ is 0.9375 itself. The nearest central is 1.40625 and the difference is 0.46875. Comparing the power-base for both, 90/1.40625=64 and 60/0.46875=128. If we reduce the power-base as 32, 90/32= 2.8125 and 60/32=1.875 and difference is intended angle of 0.9375. Therefore, appropriate set of angle is 2.8125+60/32.

Taking the Angle-Rewritten for 2.8125 +0.9375;

0.9375= 45 - (22.5 + (11.25 + (5.625 + (2.8125-0.9375]

Sin 0.9375°= $\sqrt{(2 - \sqrt{(2 + \sqrt{(2 + \sqrt{(2 + \sqrt{(2 - \sqrt{3})})})})}]/2}$

Ex-iii. For Sin 8.4375°, first of all this number (removing decimals) is perfectly divisible by 3. The angle does not fall in half-series. So, subtracting nearest central of half-series is 8.4375 - 5.625 = 2.8125 which is half-series for route-90.

Applying Angle-Rewritten method for 5.625, it will be 45 - (22.5 + (11.25 + (5.625]. Adjusting reminder of 2.8125, Angle-Rewritten method for 8.4375 will be 45 - (22.5 + (11.25 + (***5.625 - 2.8125***].

Sin 8.4375°= $\sqrt{(2 - \sqrt{(2 + \sqrt{(2 + \sqrt{(2 - \sqrt{2})})})}]/2}$

Ex-iii. For Sin 4.6875°, first of all this number (removing decimals) is perfectly divisible by 3. The angle does not fall in half-series. So, subtracting nearest central of half-series is 4.6875 - 2.8125 = 1.875 which is half-series for route-60.

In this case, applying Angle-Rewritten method for Central of 2.8125, the result will be 45 - (22.5 + (11.25 + (5.625 + **(2.8125]** with power-base 32. Adjusting the difference of 1.875 (having same power-base 32) with the central, the Angle-Rewritten will be 45 - (22.5 + (11.25 + (5.625 + **(2.8125 - 1.875]**. Therefore, Sin 4.6875° = $\sqrt{(2 - \sqrt{(2 + \sqrt{(2 + \sqrt{(2 - \sqrt{3})})})}]/2}$

EXACT VALUES IN TRIGONOMETRY

If we use Binary-sign technique, the solution will be:

	Given angle	4.6875	
	60/32	1.875	-ve
	Centrals	2.8125	
	45	45	2
	-22.5	22.5	-2
	-11.25	11.25	2
	-5.625	5.625	2
	-2.8125	2.8125	2

Sin 4.6875° = √(2 - √(2 + √(2 + √(2 + √(2 - √3]/2

Ex-iv. For **Sin 11.8125°**, first of all this number (removing decimals) is perfectly divisible by 3. The angle does not fall in half-series. Therefore, subtracting nearest central, 11.8125 - 11.25 = 0.5625 which is half-series for route-36° with power-base of 64 (=36/0.5625). In the central of 11.25, power-base is 8. To make same power-base as of central, we should make 0.5625 in power-base 8. Taking chords (a and b) of 36°, power-base 8 for 0.5625° will be:

Angle		A	B
36		√(2 -(√(6+2√5)/2]	√(10+2√5)/2
18		√(2 -√(10+2√5)/2)	√(2 +√(10+2√5)/2)
9		√(2 -√(2 +√(10+2√5)/2]	√(2 + √(2 +√(10+2√5)/2]
4.5	1	√(2-√(2+√(2+√(10+2√5)/2]	√(2+√(2+√(2+√(10 + 2√5)/2]
2.25	2		
1.125	4		
0.5625	8		

a= √(2 -√(2 + √(2 +√(10+2√5)/2];

b=√(2 + √(2 + √(2 +√(10+2√5)/2]; and

ab=√(2 -√(2 +√(10+2√5)/2]

In this case, applying Angle-Rewritten method for Central of 11.25, the result will be 45 - (22.5 + (**11.25**].

ANGLE-REWRITTEN METHOD

Adjusting the difference of 0.5625 with the central, the Angle-Rewritten will be 45 - (22.5 + (***11.25 - 0.5625***].

Sin 11.8125° = √(2 - √(2 + √(*2 - √(2 - √(2 + √(10 + 2√5)/2))*]/2.

Ex-v. Let us take example of Sin 15.9375°. The angle is divisible by 3 but not in half-series. We need nearest central for this angle, which is 11.25 with reminder 4.6875. The result is not in half-series but we can subtract 2.8125 from this getting 1.875 within half-series of route-60 having power-base of 32 (=60/1.875). Subtracting this reminder of 1.875 from given angle 15.9375, new central is 14.0625. The power-base is 32 which is same as in 1.875, so, we our formula has dependency of ab (1×√3=√3).

		A			15.9375
	Reminder (Route-60)				1.875
	Angle-Rewritten for				14.0625
0		90	45	45	45
0		45	22.5	-22.5	22.5
0		22.5	11.25	-11.25	11.25
0		11.25	5.625	5.625	16.875
0		5.625	2.8125	-2.8125	14.0625
14.0625	=45-(22.5+(11.25-(5.625-(***2.8125***]				
15.9375	=45-(22.5+(11.25-(5.625-(***2.8125-1.875***]				
Sin 15.9375°	=√(2 -√(2 + √(2 -√(2 -√(2 - √*3*]/2				

Ex-vi. Let us take one example in radian-base of Sin 1.390809. The angle has given in radian. For this either we need conversion to degrees or need Half-series in radians. Central angle near to 1.390809ᶜ would be:

			Central		Combined Half	
				1.0471976	(-)	
A	-	1.570796	0.785398	0.523599	0.261799	√(2 +
A/2	1.570796	0.785398	1.178097	0.261799	0.916298	√(2 +
A/4	1.570796	1.178097	1.374447	0.130900	1.243547	√(2 +

EXACT VALUES IN TRIGONOMETRY

| A/8 | 1.570796 | 1.374447 | 1.472622 | 0.065450 | 1.407172 | √(2 − |
| A/16 | 1.374447 | 1.472622 | 1.423534 | 0.032725 | **1.390809** | √(2 +ab) |

Without any explanation,

Sin 1.390809 = √(2 + √(2 + √(2 + √(2 − √(2 + ab]/2

= √(2 + √(2 + √(2 + √(2 − √(2 + √3]/2

Ex-vii. Let us take one example in grade-system with Sine in each half angle having binary-sign technique without any explanation.

Grade	A	Binary Sign	Sin A
100	100		2/2
50	50	2	√2]/2
−25	25	−2	√(2 −√2]/2
−12.5	12.5	2	√(2 −√(2 +√2]/2
−6.25	6.25	2	√(2 −√(2 +√(2 +√2]/2
−3.125	3.125	2	√(2 −√(2 +√(2 +√(2 +√2]/2
−1.5625	1.5625	2	√(2 −√(2 +√(2 +√(2 +√(2 +√2]/2
−0.78125	0.78125	2	√(2 −√(2 +√(2 +√(2 +√(2 +√2]/2

Other trigonometric relations than Sine of an angle

In the above process, we obtained chord (a) for an angle. If we change the sign of first radical, the result is supplementary chord (b). For example, chord (a) for double of 42.1875ᵍ is √(2 −√(2 −√(2 +√(2 −√(2 −√2]. Changing the sign of first radical, √(2 +√(2 −√(2 +√(2 −√(2 −√2] will be the supplementary chord (b) for double of 42.1875ᵍ.

Once, we have chords (a, b, and ab), we can calculate all trigonometric ratios (Cos A, Tan A etc.) using formulae for each trigonometric relation given on page 2.

Chapter 5
PRECISE-REWRITTEN METHOD

KEY WORDS – Precise-Rewritten, appropriate accuracy, irrational number, binary-sign technique.

5.1 Precise-Rewritten Concept

As discussed earlier, only the angle in degrees that is divisible by 3 and last two digits after decimal are divisible by 25 has exact value of Sine. Even we say that exact value of Sine of an angle; in fact, there is no exact value of Sine except for 0°, 30° and 90°. All the remaining cases of ultimate value of Sine of an angle is always irrational.

For example, well-known value for Sin 60° is √3/2 under radical notation. It seems as exact value, but in real number system (decimal system), we should relay on the precision of digits after decimals, because √3 itself is an irrational number[14]. Similar the case for Sin 45° or 36°. All these angles, which have so-called exact value of trigonometric ratios, have a value near to exact in decimal system.

In other words, in the decimal system of number, trigonometric ratios of an angle is that number which is *appropriate accuracy* for the user. Under this concept of *appropriate accuracy*, we can calculate exact value of any angle using *Precise-Rewritten method*.

5.1.1 Inputs

For Precise-Rewritten method, we need three generic inputs: degree measure of angle (we can change it into another measure too), *appropriate accuracy* of precision (say 10-digits accurate, 20-digits accurate, or n-digits accurate) and half-series under binary-sign technique.

Based on above three inputs, we can compute trigonometric ratios of *any angle* using the simple arithmetical process. After few exercise, student at elementary school who has general knowledge of trigonometry can find the exact trigonometric value for any angle.

The result before dividing by 2, is chord (a) for the given number. From this chord (a), we can compute supplementary chord (b) easily. From 'a', 'b' and 'ab', we are able to compute any trigonometric ratio. Need not to say, we have computed only the Sine of an angle as representation of trigonometric ratios in this book.

---

[14] Guru Theodorus (5th Century BC) proved √3 is an irrational. For his respect, it is called as Theodorus' constant.

5.1.2 Accuracy

For *any angle*, accuracy under Precise-Rewritten method is based on the user-defined precision level. For high accuracy, more arithmetical process requires. Using Precise-Rewritten method, Sin 1° to Sin 45° has given with exact value having accuracy at 15-digits after decimal in the end of this Chapter.

As already explained, either √2, √3 or √5 is an irrational number. For practical purpose, we should set some digits after decimal for √2, √3 or √5. Recommended appropriate accuracy under Precise-Rewritten method may be same number of digits of accuracy we accept for √2, √3 or √5. Of course, we can set higher accuracy too.

For example, we may use 1.732050807568880 for √3 and 0.866025403784439 for √3/2 having 15-digits after decimal point. Carefully observing on these two figures, former is not exactly double of the later. This is the *appropriate accuracy* accepted by the user in case of √3.

Under Precise-Rewritten method, Sin 60° is √(2 + √(2 -√(2 -√(2 -√(2 -√(2 -√(2 -√(2 -√(2 -√(2 -√(2 -√(2 -√(2 -√(2 -√(2 -√(2 -√(2 -√(2 -√(2 -√(2 -√(2 -√(2 -√(2 -√(2 -√(2 -√(2 -√(2 -√(2 -√(2 -√(2 -√(2 -√(2 -√(2 -√(2 -√(2 -√(2 -√(2 -√(2 -√(2 -√(2 -√2]/2. Its decimal value[15] is 0.866025403784439, which is same as above[16].

---

[15] For an infinitive series with nested radicals for 2 in form of √(2 -√(2 - ……. is equal to 1 and √(2 +√(2 +…. is equal to 2. In the above case Sin 60° has value of √(2 + √(2 -√(2 - …]/2, which is, in fact, √(2 +1)/2 = √3/2. Therefore, for an infinitive series under Precise-Rewritten method, trigonometric ratios will be exact value.

[16] To make the presentation easy, all the closing brackets are collapsed by single bracket ']' in this Chapter.

Another example of 20/3 is 0.6666666_6.... Multiplying back with 3, the exact value is something other than 20. Similar the case with 1/3 and back multiplication by 3. In these cases, appropriate accuracy for 1 is 0.99999_9 ... and vice versa. But for practical purpose, we cannot say an infinitive digits, rather we set a fix-length; e.g. 15-digits etc. In that case, we assumed 15 numbers of 9's after decimal is 1. This is the user-defined *appropriate accuracy* for our purpose.

Appropriate accuracy have two input parameters. The first is the number of digits handled during each process of calculation and the second is the number of calculations itself. If we define the appropriate accuracy of 15-digits, we must use, *at least,* 15-dgits after decimal in each step of calculation. When we arrive at the 15-digits after decimal, our maximum error will be half of remainder in 15th digit[17].

For example, if we set 15-digits accuracy, we should take *at least* 15-digits during calculation. The difference after those 15-digits may be maximum of half of 15th digit; e.g. 23.4±0.0000000000000005 for 23.4. However, there were numerous '+' or '−' within our calculation, the difference is less ±0.0000000000000005.

To reduce this gap, of course, we should increase the calculation process vis-à-vis no. of digits in each step. Roughly, 10 further calculations increases the appropriate accuracy by 3-digits. Therefore, to obtain accuracy of 100-digits, we should handle 100-digits after decimal in each step and have around 300 nested radicals.

---

[17] Using, sum of series of powers for $1/n = \frac{1}{n-1}$ for n=2; our precision will have maximum difference of 0.5. This maximum will occur, if we subtract or add the halves infinitively. If we have used + or − sign during calculations, the differences fall less than half of last digit.

5.1.3 Mathematical process

In this method, one need to perform just simple arithmetical process of Angle-Rewritten for route-90° until achieving *appropriate accuracy*. In each step of process, it gives the chord (a) of corresponding angle. Since our recommended route is route-90, we need not determine supplementary chord (b).

Let's take an example of Sin 1° for explaining the method along with example.

- Make a table with 3 columns:
  - First column contains *90 and its half in each step.* The sign has to be assigned based on target angle (1° in our case). If the earlier cumulative value (in second column) is more than target 1°, the sign will be –ve; if the earlier cumulative value is less than the target, the sign will be +ve.
  - Second column is cumulative angle. We should care the cumulative value, whether achieved at *appropriate accuracy* or not. Any digits after precise no. of digits accuracy may remove or round.
  - Third column is the ratio of first column and must be 2 in each steps. Its *sign is critical* for Sin A. In our case, 90/45= 2 is first sign; 45/(-22.5)= -2 is second sign and so on.

		Sign
90.00000000000000		
45.00000000000000	45.00000000000000	2
-22.50000000000000	22.50000000000000	-2
-11.25000000000000	11.25000000000000	2
-5.62500000000000	5.62500000000000	2
-2.81250000000000	2.81250000000000	2
-1.40625000000000	1.40625000000000	2
-0.70312500000000	0.70312500000000	2
0.35156250000000	1.05468750000000	-2
-0.17578125000000	0.87890625000000	-2

EXACT VALUES IN TRIGONOMETRY

0.08789062500000	0.96679687500000	-2
0.04394531250000	1.01074218750000	2
-0.02197265625000	0.98876953125000	-2
0.01098632812500	0.99975585937500	-2
0.00549316406250	1.00524902343750	2
-0.00274658203125	1.00250244140625	-2
-0.00137329101563	1.00112915039062	2
-0.00068664550781	1.00044250488281	2
-0.00034332275391	1.00009918212890	2
-0.00017166137695	0.99992752075195	2
0.00008583068848	1.00001335144042	-2
-0.00004291534424	0.99997043609619	-2
0.00002145767212	0.99999189376831	-2
0.00001072883606	1.00000262260437	2
-0.00000536441803	0.99999725818634	-2
0.00000268220901	0.99999994039536	-2
0.00000134110451	1.00000128149986	2
-0.00000067055225	1.00000061094760	-2
-0.00000033527613	1.00000027567148	2
-0.00000016763806	1.00000010803341	2
-0.00000008381903	1.00000002421438	2
-0.00000004190952	0.99999998230487	2
0.00000002095476	1.00000000325962	-2
-0.00000001047738	0.99999999278225	-2
0.00000000523869	0.99999999802094	-2
0.00000000261934	1.00000000064028	2
-0.00000000130967	0.99999999933061	-2
0.00000000065484	0.99999999998545	-2
0.00000000032742	1.00000000031287	2
-0.00000000016371	1.00000000014916	-2
-0.00000000008185	1.00000000006730	2
-0.00000000004093	1.00000000002638	2
-0.00000000002046	1.00000000000591	2
-0.00000000001023	0.99999999999568	2
0.00000000000512	1.00000000000080	-2
-0.00000000000256	0.99999999999824	-2

PRECISE-REWRITTEN METHOD

0.00000000000128	0.99999999999952	-2
0.00000000000064	1.00000000000016	2
-0.00000000000032	0.99999999999984	-2
0.00000000000016	1.000000000000000	-2

For writing Sine of target angle, start from first 2 (always +ve). Each of 2's should be written as $\sqrt{(}2$. The sign before this $\sqrt{(}2$ should be same as above from the table. In our case

Sin 1° = $\sqrt{(}2$ -$\sqrt{(}2$ + $\sqrt{(}2$ + $\sqrt{(}2$ + $\sqrt{(}2$ + $\sqrt{(}2$ + $\sqrt{(}2$ -$\sqrt{(}2$ -$\sqrt{(}2$ -$\sqrt{(}2$ + $\sqrt{(}2$ -$\sqrt{(}2$ -$\sqrt{(}2$ + $\sqrt{(}2$ -$\sqrt{(}2$ + $\sqrt{(}2$ + $\sqrt{(}2$ + $\sqrt{(}2$ + $\sqrt{(}2$ -$\sqrt{(}2$ -$\sqrt{(}2$ - $\sqrt{(}2$ + $\sqrt{(}2$ -$\sqrt{(}2$ -$\sqrt{(}2$ + $\sqrt{(}2$ -$\sqrt{(}2$ + $\sqrt{(}2$ + $\sqrt{(}2$ + $\sqrt{(}2$ + $\sqrt{(}2$ -$\sqrt{(}2$ - $\sqrt{(}2$ -$\sqrt{(}2$ + $\sqrt{(}2$ -$\sqrt{(}2$ -$\sqrt{(}2$ + $\sqrt{(}2$ -$\sqrt{(}2$ + $\sqrt{(}2$ + $\sqrt{(}2$ + $\sqrt{(}2$ - $\sqrt{(}2$ -$\sqrt{(}2$ -$\sqrt{(}2$ + $\sqrt{(}2$ -$\sqrt{(}2$ -$\sqrt{}2$]/2

Since the angle is 1 accurate for 15-digits of decimal, the accuracy of Sin 1° is same as 15-digits of decimal.

Note:

1. *The figure after 15 digits has not shown in the table, but has taken into calculation base. Therefore, calculation of each step, figure may not match exactly in this and Angle-Rewritten method Chapter.*
2. *The technique to exhibit numbers here is Binary-Sign technique as described in Angle-Rewritten method. Even recommended, being handling sign is extremely easy under Binary-sign technique; readers may use other techniques as well.*
3. *The result before dividing by 2 is chord (a) of 2°. Supplementary chord (b) for 2° shall be exactly same as 'a' replacing first sign '-' by '+'.*

Cos 89° or for complementary angle, Sin 89° = $\sqrt{(}2$ +$\sqrt{(}2$ + $\sqrt{(}2$ + $\sqrt{(}2$ + $\sqrt{(}2$ + $\sqrt{(}2$ + $\sqrt{(}2$ -$\sqrt{(}2$ -$\sqrt{(}2$ -$\sqrt{(}2$ + $\sqrt{(}2$ -$\sqrt{(}2$ -$\sqrt{(}2$ + $\sqrt{(}2$ -$\sqrt{(}2$ + $\sqrt{(}2$ + $\sqrt{(}2$ + $\sqrt{(}2$ + $\sqrt{(}2$ -$\sqrt{(}2$ -$\sqrt{(}2$ - $\sqrt{(}2$ + $\sqrt{(}2$ -$\sqrt{(}2$ -$\sqrt{(}2$ + $\sqrt{(}2$ -$\sqrt{(}2$ + $\sqrt{(}2$ + $\sqrt{(}2$ + $\sqrt{(}2$ + $\sqrt{(}2$ -$\sqrt{(}2$ -$\sqrt{(}2$ -$\sqrt{(}2$ + $\sqrt{(}2$ -$\sqrt{(}2$ -$\sqrt{(}2$ + $\sqrt{(}2$ -$\sqrt{(}2$ + $\sqrt{(}2$ + $\sqrt{(}2$ + $\sqrt{(}2$ + $\sqrt{(}2$ -$\sqrt{(}2$ -$\sqrt{(}2$ + $\sqrt{(}2$ -$\sqrt{}2$]/2

EXACT VALUES IN TRIGONOMETRY

Furthermore, let us take another example of Sin 64°.

		Sign
90.00000000000000		
45.00000000000000	45.00000000000000	2
22.50000000000000	67.50000000000000	2
-11.25000000000000	56.25000000000000	-2
5.62500000000000	61.87500000000000	-2
2.81250000000000	64.68750000000000	2
-1.40625000000000	63.28125000000000	-2
0.70312500000000	63.98437500000000	-2
…….	…….	….
32-steps	omitted	
…….	…….	….
0.00000000008185	63.99999999996910	-2
0.00000000004093	64.00000000001000	2
-0.00000000002046	63.99999999998950	-2
0.00000000001023	63.99999999999980	-2
0.00000000000512	64.00000000000490	2
-0.00000000000256	64.00000000000230	-2
-0.00000000000128	64.00000000000110	2
-0.00000000000064	64.00000000000040	2
-0.00000000000032	64.00000000000010	2
-0.00000000000016	63.99999999999990	2
0.00000000000008	64.000000000000000	-2

So, Sin 64°= √(2 + √(2 -√(2 -√(2 + √(2 -√(2 -√(2 + √(2 -√(2 + √(2 + √(2 + √(2 + √(2 -√(2 -√(2 -√(2 + √(2 -√(2 -√(2 + √(2 -√(2 + √(2 + √(2 + √(2 + √(2 -√(2 -√(2 -√(2 + √(2 -√(2 -√(2 + √(2 -√(2 + √(2 + √(2 + √(2 + √(2 -√(2 -√(2 -√(2 + √(2 -√(2 -√(2 + √(2 -√(2 + √(2 + √(2 + √(2 + √(2 -√2]/2

Cos 64° or complementary angle, Sin 26°= √(2 - √(2 -√(2 -√(2 + √(2 -√(2 -√(2 + √(2 -√(2 + √(2 + √(2 + √(2 + √(2 -√(2 -√(2 -√(2 + √(2 -√(2 -√(2 + √(2 -√(2 + √(2 + √(2 + √(2 + √(2 -√(2 -√(2 -√(2 + √(2 -√(2 -√(2 + √(2 -√(2 + √(2 + √(2 + √(2 + √(2 -√(2 -√(2 -√(2 + √(2 -√(2 -√(2 + √(2 -√(2 + √(2 + √(2 + √(2 + √(2 -√2]/2

PRECISE-REWRITTEN METHOD

Let us take an example having decimal like Sin 23.4°:

		Sign
90.00000000000000		
45.00000000000000	45.00000000000000	2
-22.50000000000000	22.50000000000000	-2
11.25000000000000	33.75000000000000	-2
-5.62500000000000	28.12500000000000	-2
-2.81250000000000	25.31250000000000	2
-1.40625000000000	23.90625000000000	2
-0.70312500000000	23.20312500000000	2
0.35156250000000	23.55468750000000	-2
-0.17578125000000	23.37890625000000	-2
....	....	....
32-steps	omitted	
....	....	....
0.00000000002046	23.39999999999920	2
0.00000000001023	23.40000000000940	2
-0.00000000000512	23.40000000000430	-2
-0.00000000000256	23.40000000000170	2
-0.00000000000128	23.40000000000050	2
-0.00000000000064	23.39999999999980	2
0.00000000000032	23.40000000000010	-2
-0.00000000000016	23.40000000000000	-2

Hence, Sin 23.4°= √(2 -√(2 -√(2 -√(2 +√(2 +√(2 +√(2 -√(2 -√(2 -√(2 +√(2 +√(2 -√(2 +√(2 +√(2 -√(2 -√(2 -√(2 -√(2 +√(2 +√(2 -√(2 +√(2 +√(2 +√(2 -√(2 -√(2 -√(2 -√(2 + √(2 +√(2 -√(2 +√(2 +√(2 +√(2 -√(2 -√(2 -√(2 +√(2 + √(2 -√(2 +√(2 +√(2 -√(2 -√2]/2

Cos 23.4° or for complementary angle, Sin 66.6°= √(2 +√(2 - √(2 -√(2 +√(2 +√(2 +√(2 -√(2 -√(2 -√(2 -√(2 +√(2 +√(2 -√(2 + √(2 +√(2 +√(2 -√(2 -√(2 -√(2 -√(2 +√(2 +√(2 -√(2 +√(2 + √(2 +√(2 -√(2 -√(2 -√(2 -√(2 +√(2 +√(2 -√(2 +√(2 +√(2 + √(2 -√(2 -√(2 -√(2 -√(2 +√(2 +√(2 -√(2 +√(2 +√(2 +√(2 -√(2 -√2]/2

EXACT VALUES IN TRIGONOMETRY

5.2 Sine of any angle

As discussed earlier and examples above, for an *appropriate accuracy*, this method gives *exact value of trigonometry ratios of any angle*, integer or decimal. In case our measurement of angle in grade or in radian, we should care the binary numbers only (not the process or result). Angles that have chord expressed on √2 is, *theoretically*, constructible.

Following table has shown exact value of Sine of an angle from 1° to 45° (Cosine 89° - 45°) with appropriate accuracy of 15-digits after decimal.

A°	Sin A°
1	√(2 -√(2 + √(2 + √(2 + √(2 + √(2 -√(2 -√(2 -√(2 + √(2 -√(2 -√(2 + √(2 -√(2 + √(2 + √(2 + √(2 -√(2 -√(2 -√(2 + √(2 -√(2 -√(2 + √(2 -√(2 + √(2 + √(2 + √(2 -√(2 -√(2 -√(2 + √(2 -√(2 -√(2 + √(2 -√(2 + √(2 + √(2 + √(2 -√(2 -√(2 -√(2 + √(2 -√(2 -√(2]/2
2	√(2 -√(2 + √(2 + √(2 + √(2 -√(2 -√(2 -√(2 + √(2 -√(2 -√(2 + √(2 -√(2 + √(2 + √(2 + √(2 -√(2 -√(2 -√(2 + √(2 -√(2 -√(2 + √(2 -√(2 + √(2 + √(2 + √(2 -√(2 -√(2 -√(2 + √(2 -√(2 -√(2 + √(2 -√(2 + √(2 + √(2 + √(2 -√(2 -√(2 -√(2 + √(2 -√(2 -√(2 + √(2 -√(2 + √(2 + √(2 + √(2 + √(2]/2
3	√(2 -√(2 + √(2 + √(2 + √(2 -√(2 -√(2 + √(2 + √(2 -√(2 -√(2 + √(2 + √(2 -√(2 -√(2 + √(2 + √(2 -√(2 -√(2 + √(2 + √(2 -√(2 -√(2 + √(2 + √(2 -√(2 -√(2 + √(2 + √(2 -√(2 -√(2 + √(2 + √(2 -√(2 -√(2 + √(2 + √(2 -√(2 -√(2 + √(2 + √(2 -√(2 -√(2 + √(2 + √(2]/2
4	√(2 -√(2 + √(2 + √(2 -√(2 -√(2 -√(2 + √(2 -√(2 -√(2 + √(2 -√(2 + √(2 + √(2 + √(2 -√(2 -√(2 -√(2 + √(2 -√(2 -√(2 + √(2 -√(2 + √(2 + √(2 -√(2 -√(2 -√(2 + √(2 -√(2 -√(2 + √(2 -√(2 + √(2 + √(2 -√(2 -√(2 + √(2 -√(2 -√(2 + √(2 -√(2 + √(2 + √(2]/2
5	√(2 -√(2 + √(2 + √(2 -√(2 + √(2 + √(2 -√(2 + √(2 + √(2 -√(2 + √(2 -√(2 + √(2 + √(2 -√(2 + √(2 + √(2 -√(2 + √(2 + √(2 -√(2 + √(2 + √(2 -√(2 + √(2 + √(2 -√(2 + √(2 + √(2 -√(2 + √(2 + √(2 -√(2 + √(2 -√(2 + √(2 + √(2 -√(2 + √(2 + √(2 -√(2 + √(2 + √(2 -√(2 + √2]/2

PRECISE-REWRITTEN METHOD

6 √(2 -√(2 + √(2 + √(2 -√(2 -√(2 + √(2 + √(2 -√(2 -√(2 + √(2 + √(2 -√(2 -√(2 + √(2 + √(2 -√(2 -√(2 + √(2 + √(2 -√(2 -√(2 + √(2 + √(2 -√(2 -√(2 + √(2 + √(2 -√(2 -√(2 + √(2 + √(2 -√(2 -√(2 + √(2 + √(2 -√(2 -√(2 + √(2 + √2]/2

7 √(2 -√(2 + √(2 + √(2 -√(2 -√(2 + √(2 -√(2 + √(2 + √(2 + √(2 + √(2 -√(2 -√(2 -√(2 + √(2 -√(2 -√(2 + √(2 -√(2 + √(2 + √(2 + √(2 + √(2 -√(2 -√(2 -√(2 + √(2 -√(2 -√(2 + √(2 -√(2 + √(2 + √(2 + √(2 -√(2 -√(2 -√(2 + √(2 -√(2 -√(2 + √(2 -√(2 + √(2 + √(2 + √(2 -√(2 -√(2 -√(2 + √(2 -√2]/2

8 √(2 -√(2 + √(2 + √(2 -√(2 -√(2 -√(2 + √(2 -√(2 -√(2 + √(2 -√(2 + √(2 + √(2 + √(2 + √(2 -√(2 -√(2 -√(2 + √(2 -√(2 -√(2 + √(2 -√(2 + √(2 + √(2 + √(2 + √(2 -√(2 -√(2 -√(2 + √(2 -√(2 -√(2 + √(2 -√(2 + √(2 + √(2 + √(2 + √(2 -√(2 -√(2 -√(2 + √(2 -√(2 -√(2 + √(2 -√(2 + √(2 + √(2 + √(2 -√2]/2

9 √(2 -√(2 + √(2 + √(2 -√2]/2

10 √(2 -√(2 + √(2 + √(2 -√(2 + √(2 + √(2 -√(2 + √(2 + √(2 -√(2 + √(2 + √(2 -√(2 + √(2 + √(2 -√(2 + √(2 + √(2 -√(2 + √(2 + √(2 -√(2 + √(2 + √(2 -√(2 + √(2 + √(2 -√(2 + √(2 + √(2 -√(2 + √(2 + √(2 -√(2 + √(2 + √(2 -√(2 + √(2 + √(2 -√(2 + √(2 + √(2 -√(2 + √(2 + √(2 -√(2 + √(2 + √(2 -√(2 + √(2 + √(2 -√2]/2

11 √(2 -√(2 + √(2 + √(2 -√(2 + √(2 + √(2 + √(2 + √(2 -√(2 -√(2 -√(2 + √(2 -√(2 -√(2 + √(2 -√(2 + √(2 + √(2 + √(2 + √(2 -√(2 -√(2 -√(2 + √(2 -√(2 -√(2 + √(2 -√(2 + √(2 + √(2 + √(2 + √(2 -√(2 -√(2 -√(2 + √(2 -√(2 -√(2 + √(2 + √(2 + √(2 + √(2 -√(2 -√(2 -√(2 + √2]/2

12 √(2 -√(2 + √(2 -√(2 -√(2 + √(2 + √(2 -√(2 -√(2 + √(2 + √(2 -√(2 -√(2 + √(2 + √(2 -√(2 -√(2 + √(2 + √(2 -√(2 -√(2 + √(2 + √(2 -√(2 -√(2 + √(2 + √(2 -√(2 -√(2 + √(2 + √(2 -√(2 -√(2 + √(2 + √(2 -√(2 -√(2 + √(2 + √(2 -√(2 -√(2 + √(2 + √(2 -√(2 -√(2 + √(2 + √2]/2

13 √(2 -√(2 + √(2 -√(2 -√(2 + √(2 -√(2 -√(2 + √(2 -√(2 + √(2 + √(2 + √(2 + √(2 -√(2 -√(2 -√(2 + √(2 -√(2 -√(2 + √(2 -√(2 + √(2 + √(2 + √(2 -√(2 -√(2 -√(2 + √(2 -√(2 -√(2 + √(2 -√(2 + √(2 + √(2 + √(2 -√(2 -√(2 -√(2 + √(2 -√(2 -√(2 + √(2 -√(2 + √(2 + √(2 + √2]/2

14 √(2 -√(2 + √(2 -√(2 + √(2 -√(2 + √(2 + √(2 + √(2 + √(2 -√(2 -√(2 -√(2 + √(2 -√(2 -√(2 + √(2 -√(2 + √(2 + √(2 + √(2 + √(2 -√(2 -√(2 -√(2 + √(2 -√(2 -√(2 + √(2 -√(2 + √(2 + √(2 + √(2 -√(2 -√(2 -√(2 + √(2 -√(2 -√(2 + √(2 -√(2 + √(2 + √(2 + √(2 -√2]/2

EXACT VALUES IN TRIGONOMETRY

15 √(2 -√(2 + √(2 -√(2 -√(2 -√(2 -√(2 -√(2 -√(2 -√(2 -√(2 -√(2 -√(2 -√(2 -√(2 -√(2 -√(2 -√(2 -√(2 -√(2 -√(2 -√(2 -√(2 -√(2 -√(2 -√(2 -√(2 -√(2 -√(2 -√(2 -√(2 -√(2 -√(2 -√(2 -√(2 -√(2 -√(2 -√2]/2

16 √(2 -√(2 + √(2 -√(2 -√(2 -√(2 + √(2 -√(2 -√(2 + √(2 -√(2 + √(2 + √(2 + √(2 + √(2 -√(2 -√(2 -√(2 + √(2 -√(2 -√(2 + √(2 -√(2 + √(2 + √(2 + √(2 + √(2 -√(2 -√(2 -√(2 + √(2 -√(2 -√(2 + √(2 -√(2 + √(2 + √(2 + √(2 -√(2 -√(2 -√(2 + √(2 -√(2 -√(2 + √(2 -√(2 + √(2 + √(2 + √2]/2

17 √(2 -√(2 + √(2 -√(2 + √(2 -√(2 + √(2 + √(2 + √(2 + √(2 -√(2 -√(2 -√(2 + √(2 -√(2 -√(2 + √(2 -√(2 + √(2 + √(2 + √(2 + √(2 -√(2 -√(2 -√(2 + √(2 -√(2 -√(2 + √(2 -√(2 + √(2 + √(2 + √(2 + √(2 -√(2 -√(2 -√(2 + √(2 -√(2 -√(2 + √(2 -√(2 + √(2 + √(2 + √(2 -√(2 -√(2 -√(2 + √2]/2

18 √(2 -√(2 + √(2 -√2]/2

19 √(2 -√(2 + √(2 -√(2 + √(2 -√(2 -√(2 + √(2 -√(2 + √(2 + √(2 + √(2 + √(2 -√(2 -√(2 -√(2 + √(2 -√(2 -√(2 + √(2 -√(2 + √(2 + √(2 + √(2 + √(2 -√(2 -√(2 -√(2 + √(2 -√(2 -√(2 + √(2 -√(2 + √(2 + √(2 + √(2 -√(2 -√(2 -√(2 + √(2 -√(2 -√(2 + √(2 -√(2 + √(2 + √(2 + √2]/2

20 √(2 -√(2 + √(2 -√(2 + √(2 + √(2 -√(2 + √(2 + √(2 -√(2 + √(2 + √(2 -√(2 + √(2 + √(2 -√(2 + √(2 + √(2 -√(2 + √(2 + √(2 -√(2 + √(2 + √(2 -√(2 + √(2 + √(2 -√(2 + √(2 + √(2 -√(2 + √(2 + √(2 -√(2 + √(2 + √(2 -√(2 + √(2 + √(2 -√(2 + √(2 + √(2 -√(2 + √(2 + √(2 -√(2 + √(2 + √(2 -√(2 + √2]/2

21 √(2 -√(2 + √(2 -√(2 + √(2 + √(2 -√(2 -√(2 + √(2 + √(2 -√(2 -√(2 + √(2 + √(2 -√(2 -√(2 + √(2 + √(2 -√(2 -√(2 + √(2 + √(2 -√(2 -√(2 + √(2 + √(2 -√(2 -√(2 + √(2 + √(2 -√(2 -√(2 + √(2 + √(2 -√(2 -√(2 + √(2 + √(2 + √(2 -√(2 -√(2 + √(2 + √2]/2

22 √(2 -√(2 + √(2 -√(2 + √(2 + √(2 + √(2 + √(2 -√(2 -√(2 + √(2 -√(2 -√(2 + √(2 -√(2 + √(2 + √(2 + √(2 -√(2 -√(2 -√(2 + √(2 -√(2 -√(2 + √(2 -√(2 + √(2 + √(2 + √(2 -√(2 -√(2 -√(2 + √(2 -√(2 -√(2 + √(2 -√(2 + √(2 + √(2 + √(2 -√(2 -√(2 -√(2 + √(2 -√(2 -√2]/2

23 √(2 -√(2 -√(2 -√(2 + √(2 + √(2 + √(2 + √(2 -√(2 -√(2 -√(2 + √(2 -√(2 -√(2 + √(2 -√(2 + √(2 + √(2 + √(2 + √(2 -√(2 -√(2 -√(2 + √(2 -√(2 -√(2 + √(2 -√(2 + √(2 + √(2 + √(2 -√(2 -√(2 -√(2 + √(2 -√(2 -√(2 + √(2 -√(2 + √(2 + √(2 + √(2 -√(2 -√(2 -√(2 + √(2 -√(2 -√2]/2

24 √(2 -√(2 -√(2 -√(2 + √(2 + √(2 -√(2 -√(2 + √(2 + √(2 -√(2 -√(2 + √(2 + √(2 -√(2 -√(2 + √(2 + √(2 -√(2 -√(2 + √(2 + √(2 -√(2 -√(2 + √(2 + √(2 -√(2 -√(2 + √(2 + √(2 -√(2 -√(2 + √(2 + √(2 -√(2 -√(2 + √(2 + √(2 -√(2 + √(2 + √(2 -√(2 -√(2 + √(2 + √2]/2

PRECISE-REWRITTEN METHOD

25 √(2 -√(2 -√(2 -√(2 + √(2 + √(2 -√(2 + √(2 + √(2 -√(2 + √(2 + √(2 -√(2
+ √(2 + √(2 -√(2 + √(2 + √(2 -√(2 + √(2 + √(2 -√(2 + √(2 + √(2 -√(2
+ √(2 + √(2 -√(2 + √(2 + √(2 -√(2 + √(2 + √(2 -√(2 + √(2 + √(2 -√(2
+ √(2 + √(2 -√(2 + √(2 + √(2 -√(2 + √(2 + √(2 -√(2 + √(2 + √(2 -√(2
+ √2]/2

26 √(2 -√(2 -√(2 -√(2 + √(2 -√(2 -√(2 + √(2 -√(2 + √(2 + √(2 + √(2 + √(2
-√(2 -√(2 -√(2 + √(2 -√(2 -√(2 + √(2 -√(2 + √(2 + √(2 + √(2 -√(2 -√(2
-√(2 -√(2 + √(2 -√(2 -√(2 + √(2 -√(2 + √(2 + √(2 + √(2 -√(2 -√(2 -√(2
-√(2 + √(2 -√(2 -√(2 + √(2 -√(2 + √(2 + √(2 + √(2 + √(2 -√2]/2

27 √(2 -√(2 -√(2 -√(2 + √(2 -√(2 + √(2 -√(2 + √(2 -√(2 + √(2 -√(2 + √(2 -
√(2 + √(2 -√(2 + √(2 -√(2 + √(2 -√(2 + √(2 -√(2 + √(2 -√(2 + √(2 -√(2
+ √(2 -√(2 + √(2 -√(2 + √(2 -√(2 + √(2 -√(2 + √(2 -√(2 + √(2 -√(2 +
√(2 -√(2 + √(2 -√(2 + √(2 -√(2 + √(2 -√(2 + √(2 -√(2 + √(2 -√2]/2

28 √(2 -√(2 -√(2 -√(2 + √(2 -√(2 + √(2 + √(2 + √(2 -√(2 -√(2 -√(2 +
√(2 -√(2 -√(2 + √(2 -√(2 + √(2 + √(2 + √(2 -√(2 -√(2 -√(2 + √(2
-√(2 -√(2 + √(2 -√(2 + √(2 + √(2 + √(2 -√(2 -√(2 -√(2 + √(2 -√(2
-√(2 + √(2 -√(2 + √(2 + √(2 + √(2 -√(2 -√(2 -√(2 + √2]/2

29 √(2 -√(2 -√(2 -√(2 -√(2 -√(2 + √(2 -√(2 -√(2 + √(2 -√(2 + √(2 + √(2 +
√(2 + √(2 -√(2 -√(2 -√(2 + √(2 -√(2 -√(2 + √(2 -√(2 + √(2 + √(2 + √(2
+ √(2 -√(2 -√(2 -√(2 + √(2 -√(2 -√(2 + √(2 -√(2 + √(2 + √(2 + √(2 +
√(2 -√(2 -√(2 -√(2 + √(2 -√(2 -√(2 + √(2 -√(2 + √(2 + √(2 + √2]/2

30 √(2 -√(2 -√(2 -√(2 -√(2 -√(2 -√(2 -√(2 -√(2 -√(2 -√(2 -√(2 -√(2 -√(2
-√(2 -√(2 -√(2 -√(2 -√(2 -√(2 -√(2 -√(2 -√(2 -√(2 -√(2 -√(2 -√(2 -√(2
-√(2 -√(2 -√(2 -√(2 -√(2 -√(2 -√(2 -√(2 -√(2 -√(2 -√(2 -√(2 -√(2 -√(2
-√(2 -√(2 -√(2 -√(2 -√2]/2

31 √(2 -√(2 -√(2 -√(2 -√(2 + √(2 -√(2 + √(2 + √(2 + √(2 + √(2 -√(2 -√(2 -
√(2 + √(2 -√(2 -√(2 + √(2 -√(2 + √(2 + √(2 + √(2 + √(2 -√(2 -√(2 -√(2
+ √(2 -√(2 -√(2 + √(2 -√(2 + √(2 + √(2 + √(2 + √(2 -√(2 -√(2 -√(2 +
√(2 -√(2 -√(2 + √(2 -√(2 + √(2 + √(2 + √(2 + √(2 -√2]/2

32 √(2 -√(2 -√(2 -√(2 + √(2 -√(2 -√(2 + √(2 -√(2 + √(2 + √(2 + √(2 +
√(2 -√(2 -√(2 -√(2 + √(2 -√(2 -√(2 + √(2 -√(2 + √(2 + √(2 + √(2 + √(2
-√(2 -√(2 -√(2 + √(2 -√(2 -√(2 + √(2 -√(2 + √(2 + √(2 + √(2 + √(2 -√(2
-√(2 -√(2 + √(2 -√(2 -√(2 + √(2 -√(2 + √(2 + √(2 + √2]/2

33 √(2 -√(2 -√(2 -√(2 -√(2 + √(2 -√(2 -√(2 + √(2 + √(2 -√(2 -√(2 +
√(2 + √(2 -√(2 -√(2 + √(2 + √(2 -√(2 -√(2 + √(2 + √(2 -√(2 -√(2 + √(2
+ √(2 -√(2 -√(2 + √(2 + √(2 -√(2 -√(2 + √(2 + √(2 -√(2 -√(2 + √(2 +
√(2 -√(2 -√(2 + √(2 + √(2 -√(2 -√(2 + √(2 + √2]/2

34 √(2 -√(2 -√(2 + √(2 -√(2 + √(2 + √(2 + √(2 + √(2 -√(2 -√(2 -√(2 + √(2
-√(2 -√(2 + √(2 -√(2 + √(2 + √(2 + √(2 + √(2 -√(2 -√(2 -√(2 + √(2 -√(2
-√(2 + √(2 -√(2 + √(2 + √(2 + √(2 + √(2 -√(2 -√(2 -√(2 + √(2 -√(2 -√(2
+ √(2 -√(2 + √(2 + √(2 + √(2 + √(2 -√(2 -√(2 -√(2 + √2]/2

EXACT VALUES IN TRIGONOMETRY

35   √(2 -√(2 -√(2 + √(2 -√(2 + √(2 + √(2 -√(2 + √(2 + √(2 -√(2 + √(2 +
     √(2 -√(2 +√(2 +√(2 -√(2 + √(2 + √(2 -√(2 + √(2 + √(2 -√(2 + √(2 +
     √(2 -√(2 +√(2 +√(2 -√(2 + √(2 + √(2 -√(2 + √(2 + √(2 -√(2 + √(2 +
     √(2 -√(2 +√(2 +√(2 -√(2 + √(2 + √(2 -√(2 + √(2 + √(2 -√(2 + √2]/2

36   √(2 -√(2 -√(2 + √(2 -√(2 + √(2 -√(2 + √(2 -√(2 + √(2 -√(2 + √(2 -√(2 +
     √(2 -√(2 + √(2 -√(2 + √(2 -√(2 + √(2 -√(2 + √(2 -√(2 + √(2 -√(2 + √(2
     -√(2 + √(2 -√(2 + √(2 -√(2 + √(2 -√(2 + √(2 -√(2 + √(2 -√(2 + √(2 -√(2
     + √(2 -√(2 + √(2 -√(2 + √(2 -√(2 + √(2 -√(2 + √(2 -√2]/2

37   √(2 -√(2 -√(2 + √(2 -√(2 -√(2 -√(2 + √(2 -√(2 -√(2 + √(2 -√(2 + √(2 +
     √(2 + √(2 + √(2 -√(2 -√(2 -√(2 + √(2 -√(2 -√(2 + √(2 -√(2 + √(2 + √(2
     + √(2 + √(2 -√(2 -√(2 -√(2 + √(2 -√(2 -√(2 + √(2 -√(2 + √(2 + √(2 +
     √(2 + √(2 -√(2 -√(2 -√(2 + √(2 -√(2 -√2]/2

38   √(2 -√(2 -√(2 + √(2 -√(2 -√(2 + √(2 -√(2 + √(2 + √(2 + √(2 + √(2 -√(2
     -√(2 -√(2 + √(2 -√(2 -√(2 + √(2 -√(2 + √(2 + √(2 + √(2 + √(2 -√(2 -√(2
     -√(2 + √(2 -√(2 -√(2 + √(2 -√(2 + √(2 + √(2 + √(2 + √(2 -√(2 -√(2 -√(2
     + √(2 -√(2 -√(2 + √(2 -√(2 + √(2 + √(2 + √(2 + √(2 -√2]/2

39   √(2 -√(2 -√(2 + √(2 -√(2 -√(2 + √(2 + √(2 -√(2 -√(2 + √(2 + √(2 -√(2 -
     √(2 + √(2 + √(2 -√(2 -√(2 + √(2 + √(2 -√(2 -√(2 + √(2 + √(2 -√(2 -√(2
     + √(2 + √(2 -√(2 -√(2 + √(2 + √(2 -√(2 -√(2 + √(2 + √(2 -√(2 -√(2 +
     √(2 + √(2 -√(2 -√(2 + √(2 + √(2 -√(2 -√(2 + √(2 + √2]/2

40   √(2 -√(2 -√(2 + √(2 + √(2 -√(2 + √(2 + √(2 -√(2 + √(2 + √(2 -√(2 +
     √(2+√(2 -√(2 + √(2 + √(2 -√(2 + √(2 + √(2 -√(2 + √(2 + √(2 -√(2 +
     √(2+√(2 -√(2 + √(2 + √(2 -√(2 + √(2 + √(2 -√(2 + √(2 + √(2 -√(2 +
     √(2+√(2 -√(2 + √(2 + √(2 -√(2 + √(2 + √(2 -√(2 + √(2 + √(2 -√(2 + √2
     ]/2

41   √(2 -√(2 -√(2 + √(2 + √(2 -√(2 -√(2 -√(2 + √(2 -√(2 -√(2 + √(2 -√(2 +
     √(2 + √(2 + √(2 + √(2 -√(2 -√(2 -√(2 + √(2 -√(2 -√(2 + √(2 -√(2 + √(2
     + √(2 + √(2 + √(2 -√(2 -√(2 -√(2 + √(2 -√(2 -√(2 + √(2 -√(2 + √(2 +
     √(2 + √(2 + √(2 -√(2 -√(2 -√(2 + √(2 -√(2 -√2]/2

42   √(2 -√(2 -√(2 + √(2 + √(2 -√(2 -√(2 + √(2 + √(2 -√(2 -√(2 + √(2 + √(2
     -√(2 -√(2 + √(2 + √(2 -√(2 -√(2 + √(2 + √(2 -√(2 -√(2 + √(2 + √(2 -√(2
     -√(2 + √(2 + √(2 -√(2 -√(2 + √(2 + √(2 -√(2 -√(2 + √(2 + √(2 -√(2 -√(2
     + √(2 + √(2 -√(2 -√(2 + √(2 + √(2 -√(2 -√(2 + √(2 + √2]/2

43   √(2 -√(2 -√(2 + √(2 + √(2 + √(2 -√(2 -√(2 + √(2 -√(2 -√(2 + √(2 -
     √(2 + √(2 + √(2 + √(2 + √(2 -√(2 -√(2 + √(2 -√(2 -√(2 + √(2 -√(2
     + √(2 + √(2 + √(2 + √(2 -√(2 -√(2 -√(2 + √(2 -√(2 -√(2 + √(2 -√(2 +
     √(2 + √(2 + √(2 + √(2 -√(2 -√(2 -√(2 + √(2 -√(2 -√2]/2

44   √(2 -√(2 -√(2 + √(2 + √(2 + √(2 + √(2 -√(2 -√(2 -√(2 + √(2 -√(2 -√(2 +
     √(2 -√(2 + √(2 + √(2 + √(2 + √(2 -√(2 -√(2 -√(2 + √(2 -√(2 -√(2 + √(2
     -√(2 + √(2 + √(2 + √(2 + √(2 -√(2 -√(2 -√(2 + √(2 -√(2 -√(2 + √(2 -√(2
     + √(2 + √(2 + √(2 + √(2 -√(2 -√(2 -√(2 + √(2 -√(2 -√2]/2

45   √2/2

78

Chapter 6
VATSA METHOD

6.1 Ratio-based Sine Table

Sine of two angles have special and fixed relations. In case we take same interval, their ratio is same. This concept has utilized in this Vatsa method of Sine Table.

6.1.1 For Sexagesimal System (Degrees)

Based on Degree of measurement of angle, Sine Table has constructed below. In this calculation, value of π has assumed as 3.14159265359. Changing the value of π for more accuracy or for estimation, value of Sine shall be different. For this method, we need two fixed inputs.

First Input: Using any one method of calculation, Sin 1º will be 0.017452406437. To make Sine-Table easy, this is one of the basic value for our purpose.

EXACT VALUES IN TRIGONOMETRY

Second Input: In each increment of 1° to the Sine-Table, next value of sine will be 1/3282.88968461374 (=0.000304609687217577) times earlier Sine. Therefore, 0.000304609687217577 is another basic value.

Using two inputs above now make a table having 4 columns.

- Column E is degrees, increased by 1 in each stage.

- Column N is product of Sine of earlier angle and 0.000304609687217577 for each degree. For 1°, this is 0.00. For 2°, this is 0.017452406437× 0.00030460968721757 = 0.000005316172066 and so on.

- Column I is $I_{earlier}$ less values in column N. For 1°, this is 0.017452406437. For 2°, this is 0.017452406437- 0.000005316172066 = 0.017447090265 and so on.

- Column S is Value of Sine of that degree. Sin 1° is 0.017452406437, Sin 2° is 0.034899496703 and so on.

E Degree	N = $S_{Earlier}$ × 0.00030460968721757	I = $I_{Earlier}$ - N	S Sin (E) = S+I
1	0.000000000000	0.017452406437	0.017452406437
2	0.000005316172066	0.017447090265	0.034899496703
3	0.000010630724775	0.017436459540	0.052335956243
4	0.000015942039261	0.017420517501	0.069756473744
5	0.000021248497649	0.017399269004	0.087155742748
6	0.000026548483538	0.017372720520	0.104528463268
7	0.000031840382501	0.017340880137	0.121869343405
8	0.000037122582576	0.017303757555	0.139173100960
9	0.000042393474753	0.017261364080	0.156434465040
10	0.000047651453466	0.017213712627	0.173648177667

VATSA METHOD

E Degree	N = S $_{Earlier}$ × 0.00030460968721757	I = I $_{Earlier}$ - N	S Sin (E) = S+I
11	0.000052894917085	0.017160817710	0.190808995377
12	0.000058122268400	0.017102695441	0.207911690818
13	0.000063331915109	0.017039363526	0.224951054344
14	0.000068522270303	0.016970841256	0.241921895600
15	0.000073691752950	0.016897149503	0.258819045103
16	0.000078838788375	0.016818310714	0.275637355817
17	0.000083961808741	0.016734348906	0.292371704723
18	0.000089059253527	0.016645289652	0.309016994375
19	0.000094129570001	0.016551160082	0.325568154457
20	0.000099171213697	0.016451988869	0.342020143326
21	0.000104182648881	0.016347806220	0.358367949545
22	0.000109162349020	0.016238643871	0.374606593416
23	0.000114108797250	0.016124535073	0.390731128489
24	0.000119020486835	0.016005514587	0.406736643076
25	0.000123895921627	0.015881618665	0.422618261741
26	0.000128733616521	0.015752885048	0.438371146789
27	0.000133532097909	0.015619352950	0.453990499740
28	0.000138289904125	0.015481063046	0.469471562786
29	0.000143005585898	0.015338057460	0.484809620246
30	0.000147677706783	0.015190379754	0.500000000000
31	0.000152304843609	0.015038074910	0.515038074910
32	0.000156885586904	0.014881189323	0.529919264233
33	0.000161418541329	0.014719770782	0.544639035015
34	0.000165902326102	0.014553868456	0.559192903471
35	0.000170335575421	0.014383532880	0.573576436351

EXACT VALUES IN TRIGONOMETRY

E Degree	N = S $_{Earlier}$ × 0.00030460968721757	I = I $_{Earlier}$ - N	S Sin (E) = S+I
36	0.000174716938872	0.014208815941	0.587785252293
37	0.000179045081852	0.014029770860	0.601815023152
38	0.000183318685965	0.013846452174	0.615661475326
39	0.000187536449431	0.013658915724	0.629320391050
40	0.000191697087477	0.013467218637	0.642787609687
41	0.000195799332734	0.013271419304	0.656059028991
42	0.000199841935617	0.013071577368	0.669130606359
43	0.000203823664711	0.012867753704	0.681998360063
44	0.000207743307142	0.012660010396	0.694658370459
45	0.000211599668949	0.012448410728	0.707106781187
46	0.000215391575447	0.012233019152	0.719339800339
47	0.000219117871584	0.012013901281	0.731353701619
48	0.000222777422296	0.011791123858	0.743144825477
49	0.000226369112846	0.011564754745	0.754709580223
50	0.000229891849172	0.011334862896	0.766044443119
51	0.000233344558213	0.011101518338	0.777145961457
52	0.000236726188242	0.010864792150	0.788010753607
53	0.000240035709180	0.010624756441	0.798635510047
54	0.000243272112916	0.010381484328	0.809016994375
55	0.000246434413610	0.010135049914	0.819152044289
56	0.000249521647995	0.009885528266	0.829037572555
57	0.000252532875668	0.009632995390	0.838670567945
58	0.000255467179380	0.009377528211	0.848048096156
59	0.000258323665316	0.009119204546	0.857167300702
60	0.000261101463360	0.008858103082	0.866025403784

VATSA METHOD

E Degree	N = S $_{Earlier}$ × 0.00030460968721757	I = I $_{Earlier}$ - N	S Sin (E) = S+I
61	0.000263799727369	0.008594303355	0.874619707139
62	0.000266417635426	0.008327885720	0.882947592859
63	0.000268954390090	0.008058931329	0.891006524188
64	0.000271409218642	0.007787522111	0.898794046299
65	0.000273781373316	0.007513740737	0.906307787037
66	0.000276070131532	0.007237670606	0.913545457643
67	0.000278274796112	0.006959395810	0.920504853452
68	0.000280394695492	0.006679001114	0.927183854567
69	0.000282429183933	0.006396571930	0.933580426497
70	0.000284377641708	0.006112194289	0.939692620786
71	0.000286239475298	0.005825954813	0.945518575599
72	0.000288014117572	0.005537940696	0.951056516295
73	0.000289701027955	0.005248239668	0.956304755963
74	0.000291299692599	0.004956939975	0.961261695938
75	0.000292809624534	0.004664130351	0.965925826289
76	0.000294230363821	0.004369899987	0.970295726276
77	0.000295561477689	0.004074338509	0.974370064785
78	0.000296802560668	0.003777535949	0.978147600734
79	0.000297953234712	0.003479582714	0.981627183448
80	0.000299013149314	0.003180569565	0.984807753012
81	0.000299981981614	0.002880587583	0.987688340595
82	0.000300859436497	0.002579728146	0.990268068742
83	0.000301645246681	0.002278082900	0.992546151641
84	0.000302339172800	0.001975743727	0.994521895368
85	0.000302941003479	0.001672802723	0.996194698092

EXACT VALUES IN TRIGONOMETRY

E Degree	N $= S_{Earlier} \times$ 0.00030460968721757	I $= I_{Earlier} - N$	S Sin (E) = S+I
86	0.000303450555394	0.001369352168	0.997564050260
87	0.000303867673329	0.001065484495	0.998629534755
88	0.000304192230228	0.000761292265	0.999390827019
89	0.000304424127226	0.000456868137	0.999847695156
90	0.000304563293687	0.000152304844	1.000000000000

From above table, value of Sine for each degree is easily calculative without any technical effort. General scholar may construct the table with simple knowledge of just two basic parameters (Sin 1° is 0.017452406437 and its effect with other Sine is based on it and 0.000304609687217577). To obtain high level of accuracy, we should take more digits in each basic inputs.

6.1.2 For Metric System (Grades)

Exactly same principle as degrees has adopted in the Grade system of measurement of an angle. Only the difference is the two inputs. In the Grade system, Sin 1ᵍ has assumed as 0.015707317312 from conversion process. Second the deviation from each Grade in Sine is 1/4052.93068005671 (=0.000246735036679). The Sine Table shall be as follows:

E Grade	N $= S_{-Earlier} \times$ 0.000246735036679	I $= I_{-Earlier} - N$	S Sin (E)= S+I
1		0.015707317312	0.015707317312
2	0.000003875546	0.015703441766	0.031410759078
3	0.000007750135	0.015695691632	0.047106450710
4	0.000011622812	0.015684068820	0.062790519529
5	0.000015492621	0.015668576199	0.078459095728

VATSA METHOD

E Grade	N = S-Earlier × 0.000246735036679	I = I-Earlier - N	S Sin (E)= S+I
6	0.000019358608	0.015649217591	0.094108313319
7	0.000023219818	0.015625997773	0.109734311091
8	0.000027075299	0.015598922473	0.125333233564
9	0.000030924100	0.015567998373	0.140901231938
10	0.000034765271	0.015533233103	0.156434465040
11	0.000038597863	0.015494635239	0.171929100279
12	0.000042420933	0.015452214306	0.187381314586
13	0.000046233536	0.015405980771	0.202787295357
14	0.000050034731	0.015355946040	0.218143241397
15	0.000053823581	0.015302122459	0.233445363856
16	0.000057599150	0.015244523309	0.248689887165
17	0.000061360508	0.015183162801	0.263873049965
18	0.000065106727	0.015118056074	0.278991106039
19	0.000068836881	0.015049219193	0.294040325232
20	0.000072550050	0.014976669143	0.309016994375
21	0.000076245319	0.014900423823	0.323917418198
22	0.000079921776	0.014820502047	0.338737920245
23	0.000083578513	0.014736923534	0.353474843779
24	0.000087214629	0.014649708905	0.368124552685
25	0.000090829225	0.014558879680	0.382683432365
26	0.000094421411	0.014464458270	0.397147890635
27	0.000097990299	0.014366467970	0.411514358605
28	0.000101535010	0.014264932960	0.425779291565
29	0.000105054669	0.014159878291	0.439939169856
30	0.000108548407	0.014051329884	0.453990499740
31	0.000112015363	0.013939314521	0.467929814261
32	0.000115454680	0.013823859841	0.481753674102
33	0.000118865510	0.013704994331	0.495458668432

EXACT VALUES IN TRIGONOMETRY

E Grade	N = $S_{\text{Earlier}} \times$ 0.000246735036679	I = $I_{\text{Earlier}} - N$	S Sin (E) = S+I
34	0.000122247013	0.013582747318	0.509041415750
35	0.000125598352	0.013457148966	0.522498564716
36	0.000128918703	0.013328230263	0.535826794979
37	0.000132207244	0.013196023019	0.549022817998
38	0.000135463165	0.013060559854	0.562083377852
39	0.000138685663	0.012921874191	0.575005252043
40	0.000141873942	0.012780000249	0.587785252293
41	0.000145027216	0.012634973033	0.600420225326
42	0.000148144706	0.012486828327	0.612907053653
43	0.000151225644	0.012335602683	0.625242656336
44	0.000154269270	0.012181333413	0.637423989749
45	0.000157274831	0.012024058581	0.649448048330
46	0.000160241588	0.011863816993	0.661311865324
47	0.000163168807	0.011700648186	0.673012513510
48	0.000166055767	0.011534592419	0.684547105929
49	0.000168901755	0.011365690664	0.695912796592
50	0.000171706069	0.011193984594	0.707106781187
51	0.000174468018	0.011019516577	0.718126297763
52	0.000177186918	0.010842329658	0.728968627421
53	0.000179862101	0.010662467557	0.739631094979
54	0.000182492905	0.010479974652	0.750111069631
55	0.000185078682	0.010294895970	0.760405965600
56	0.000187618794	0.010107277176	0.770513242776
57	0.000190112613	0.009917164563	0.780430407338
58	0.000192559525	0.009724605037	0.790155012376
59	0.000194958926	0.009529646111	0.799684658487
60	0.000197310224	0.009332335888	0.809016994375
61	0.000199612838	0.009132723050	0.818149717425

VATSA METHOD

E Grade	N = S-Earlier × 0.000246735036679	I = I-Earlier - N	S Sin (E)= S+I
62	0.000201866201	0.008930856850	0.827080574275
63	0.000204069756	0.008726787094	0.835807361368
64	0.000206222960	0.008520564134	0.844327925502
65	0.000208325282	0.008312238852	0.852640164354
66	0.000210376202	0.008101862650	0.860742027004
67	0.000212375216	0.007889487434	0.868631514438
68	0.000214321829	0.007675165606	0.876306680044
69	0.000216215561	0.007458950045	0.883765630089
70	0.000218055945	0.007240894100	0.891006524189
71	0.000219842527	0.007021051572	0.898027575761
72	0.000221574867	0.006799476705	0.904827052466
73	0.000223252536	0.006576224169	0.911403276636
74	0.000224875121	0.006351349049	0.917754625684
75	0.000226442221	0.006124906827	0.923879532511
76	0.000227953450	0.005896953377	0.929776485888
77	0.000229408435	0.005667544942	0.935444030830
78	0.000230806817	0.005436738124	0.940880768954
79	0.000232148251	0.005204589873	0.946085358828
80	0.000233432406	0.004971157468	0.951056516295
81	0.000234658964	0.004736498503	0.955793014799
82	0.000235827625	0.004500670879	0.960293685677
83	0.000236938098	0.004263732781	0.964557418458
84	0.000237990110	0.004025742671	0.968583161129
85	0.000238983402	0.003786759269	0.972369920398
86	0.000239917728	0.003546841541	0.975916761939
87	0.000240792858	0.003306048683	0.979222810622
88	0.000241608576	0.003064440107	0.982287250729
89	0.000242364681	0.002822075426	0.985109326155

EXACT VALUES IN TRIGONOMETRY

E Grade	N = S-Earlier × 0.000246735036679	I = I-Earlier - N	S Sin (E)= S+I
90	0.000243060986	0.002579014440	0.987688340595
91	0.000243697319	0.002335317121	0.990023657717
92	0.000244273523	0.002091043598	0.992114701315
93	0.000244789457	0.001846254141	0.993960955455
94	0.000245244993	0.001601009148	0.995561964603
95	0.000245640018	0.001355369130	0.996917333733
96	0.000245974435	0.001109394695	0.998026728429
97	0.000246248161	0.000863146534	0.998889874962
98	0.000246461130	0.000616685404	0.999506560366
99	0.000246613288	0.000370072116	0.999876632482
100	0.000246704598	0.000123367518	1.000000000000

6.2 Accuracy

Vatsa method is just arithmetical presentation of results of other method. It has just two inputs- basic angle (a) and acceleration factor (r). Its accuracy is mainly depends upon the value we take for 'a' or 'r'.

Secondly, the accuracy depends upon, how much digits in a figure we handle in a number within calculations. For example, to make a Sine Table having 50-digits of accuracy, we should have 50-digits in input 'a' or 'r'. Additionally, in each step of calculation including use of each exponentials; we should handle *at least* 50-digits.

6.3 Sine of an Individual Angle

From above, we need table for each degree or grade to determine the Sine of any number. From above ratio, we can find Sin 42° or Sin 43° in the sequence. If we need the value of Sin 44° of angle with decimal without using sequential calculation, we have to use different method.

To determine Sine of an angle, we calculate some additional steps. We can adjust these steps based on requirement of our accuracy. For gestimation figure of Sine of an angle, process will be lower with few calculation.

For individual degree, we need following parameters and formula[18]:

$r = 1/3282.88968461374 = 0.000304609687217577$

$a = 0.017452406437$ (+)

$ar = 0.017452406437/3282.88968461374$ (-)

$ar^2 = 0.017452406437/3282.88968461374^2$ (+)

$ar^3 = 0.017452406437/3282.88968461374^3$ (-)

$ar^4 = 0.017452406437/3282.88968461374^4$ (+)

$ar^5 = 0.017452406437/3282.88968461374^5$ (-)

$ar^6 = 0.017452406437/3282.88968461374^6$ (+)

$ar^7 = 0.017452406437/3282.88968461374^7$ (-)

$$n:_v = \frac{n(n+1)(n+2)(....)(n+v)}{(v+1)!}$$

For this new notation, named as *Vatsa* (see Chapter 7 Vasta-Natural Number Series for detail). Now,

---

[18] For Grade System, these values will be differ for 'a' and 'r' only. In that case r will be 1/4052.930680056710 and a will be 0.015707317312.

EXACT VALUES IN TRIGONOMETRY

$\sin 1° = ar° = 0.017452406437$

$\sin 2° = 2a - 2:_2 \, ar$

$= 2 \times 0.017452406437 - 3 \times 0.017452406437 \,/\, 3282.88968461374 = 0.034899496702$

$[\because 2:_2 = 2 \times 3/2 = 3]$

$\sin 3° = 3a - 3:_2 \, ar + 1:_4 \, ar^2.$

$= 3 \times 0.017452406437 - 6 \times 0.017452406437 \,/\, 3282.88968461374 + 1 \times 0.017452406437 \,/\, 3282.88968461374^2 = 0.05232532389796$

$\sin 4° = 4a - 4:_2 \, ar + 2:_4 \, ar^2 - 1:_6 \, ar^3$

$= 4 \times 0.017452406437 - 10 \times 0.017452406437 / 3282.88968461374 + 6 \times 0.017452406437 / 3282.88968461374^2 + 1 \times 0.017452406437 / 3282.88968461374^3$

$= 0.069756473744$

$\sin 5° = 5a - 5:_2 \, ar + 3:_4 \, ar^2 - 2:_6 \, ar^3 + 1:_8 \, ar^4$

$= 5 \times 0.017452406437 - 20 \times 0.017452406437 / 3282.88968461374 + 6 \times 0.017452406437 / 3282.88968461374^2 + 1 \times 0.017452406437 / 3282.88968461374^3$

$= 0.069756473744$

Calculation of Sine of an angle from formula up to $(n-6):_8 \, ar^4$ is almost accurate for 6 digits of decimal for angles above 5°. Using formula up to $(n-8):_{10} \, ar^5$ gives accuracy almost up to 10 digits after decimal. For more accuracy, comparatively long calculation requires as:

$\sin n° = na - n:_2 \, ar + (n-2):_4 \, ar^2 - (n-4):_6 \, ar^3 + (n-6):_8 \, ar^4 - (n-8):_{10} \, ar^5 + (n-10):_{12} \, ar^6 - (n-12):_{14} \, ar^7 +$

. . . .

Chapter 7
Vasta- Natural Number Series

7.1 New Method for Series-Sum

7.1.1 First Series (Sum of Arithmetic Series)

Mathematic scholars know the sum for natural numbers. They say this is the *sum of arithmetic series*. In case a numbers increased by or decreased by a fix deviation, the series is called as arithmetic series. Counting numbers 1, 2, 3, ... is good example of arithmetic series.

Sum for series of given natural numbers is widely usable in formula of $S_n = \frac{n}{2}(n+1)$ for natural number n (n:₁). For example:

$S_1 = \frac{1}{2}(1+1) = 1$ (So, $1_{:1} = 1$).

$S_2 = \frac{2}{2}(2+1) = 3$ (So, $2_{:1} = 3$).

$S_3 = \frac{3}{2}(3+1) = 6$ (So, $2_{:1} = 6$).

$S_4 = \frac{4}{2}(4+1) = 10$ (So, $4_{:1} = 10$).

EXACT VALUES IN TRIGONOMETRY

Hence, the sum of the series (S_n) of natural number ($n:_1$) is easy and we called it as the First Series. The list upto 10 of the First Series is:

n	$n:_1$	n	$n:_1$
1	1	6	21
2	3	7	28
3	6	8	36
4	10	9	45
5	15	10	55

On careful analysis of above formula is very simple as:

- Write the number (say 10); add 1 on it (now 10+1=11).
- Multiply both (now 10×11= 110).
- Divide it by 2 (110/2= 55).

In summary, $10:_1$= 10×11/2= 55. For other numbers as examples of the First Series:

- $20:_1$= 20×21/2= 210.
- $100:_1$= 100×101/2= 5050.
- $125:_1$= 125×126/2 = 7875.

There is new notation for sum of natural number as $1:_1$, $2:_1$, $20:_1$, $100:_1$ etc. The notation for $20:_1$ is spelt as *20 Vatsa₁* and meaning is sum of natural numbers upto 20 (i.e. 1+2+3+ …+19+20).

First Series is called as Triangular Number Sequence also. Above method is generic except new type of notation of Vatsa₁ (:₁). Vatsa has written with its initial number and common deviation as ($_a{}^d n:_1$) with meaning sum of n numbers initiating with initial number is a and common deviation is d. It has to be read as 1ˢᵗ Vatsa of a with d upto n numbers. For natural numbers notation of a and d are left being 1.

7.1.2 N_{th} Series

Taking above table again, the sum of numbers (n:$_2$) of the First Series will be as follows:

n	n:$_1$		n:$_2$
1	1	1	1
2	3	1+3	4
3	6	4+6	10
4	10	10+10	20
5	15	20+15	35
6	21	35+21	56
7	28	56+28	84
8	36	84+36	120
9	45	120+45	165
10	55	165+55	220

Finding the Second Series is another generic calculation now-a-days in the mathematics. This is the sum of sum of n natural numbers (n *Vatsa$_2$*, n:$_2$).

Similar way, we can compute the Third Series, Forth Series, Fifth Series, Sixth Series, and so on as:

N	n:$_1$	n:$_2$	n:$_3$	n:$_4$	n:$_5$	n:$_6$	n:$_7$	n:$_8$	n:$_9$	n:$_{10}$
1	1	1	1	1	1	1	1	1	1	1
2	3	4	5	6	7	8	9	10	11	12
3	6	10	15	21	28	36	45	55	66	78
4	10	20	35	56	84	120	165	220	286	364
5	15	35	70	126	210	330	495	715	1001	1365
6	21	56	126	252	462	792	1287	2002	3003	4368
7	28	84	210	462	924	1716	3003	5005	8008	12376
8	36	120	330	792	1716	3432	6435	11440	19448	31824
9	45	165	495	1287	3003	6435	12870	24310	43758	75582
10	55	220	715	2002	5005	11440	24310	48620	92378	167960

Procedure for determination of the value of any number of *Vatsa* (n:$_v$) is a simple calculation. This calculation is usable to determine value of individual *Sine of an angle*.

EXACT VALUES IN TRIGONOMETRY

Divisor: First requirement is divisor, which is just a factorial of Vatsa (n:₁) and increased accordingly:

n	n:1	n:2	n:3	n:4	n:5	n:6	n:v
1	2	3	4	5	6	7	$v+1$
1	2	6	24	120	720	5040	$(v+1)!$
	2!	3!	4!	5!	6!	7!	$(v+1)!$

For any Vatsa, divisor will be $(v+1)!$.

Required Vatsa (n:ᵥ) of given number is the second requirement. In our example, we want to 3ʳᵈ Vatsa of 9 (9:₃).

Write given number and forward numbers upto 1 more than given Vatsa (v+1). In our case, we need 9:₃, so numbers will be 9, 10, 11 and 12.

Multiply numbers in above step and divide by divisor.

$$9{:}_3 = (9\times 10\times 11\times 12)/24 = 495.$$

In the above table, 9:₃ has given as 495.

In general, the formula will be $n{:}_v = \dfrac{n(n+1)(n+2)(\ldots n+\text{upto } v+1)}{(v+1)!}$ for the first natural numbers[19]. Examples of this formula for 9 and 32:

$$9{:}_1 = \frac{9*10}{(1+1)!} = \frac{9*10}{2} = 45.$$

$$32{:}_1 = \frac{32*33}{(1+1)!} = \frac{32*33}{2} = 1056.$$

Sum of the Sum of first natural numbers is (Second Series):

$$9{:}_2 = \frac{9*10*11}{(2+1)!} = \frac{9*10*11}{6} = 165.$$

---

[19] Since this is not the intended coverage of this book to develop the method of series or their sum, above *Vatsa* are derived for the first natural numbers. These *Vatsa* are usable to find determination of Sine of an angle given in degrees.

$$32{:}_2 = \frac{32*33*34}{(2+1)!} = \frac{32*33*34}{6} = 5984.$$

Sum of the Second Series first natural numbers is (Third Series):

$$9{:}_3 = \frac{9*10*11*12}{(3+1)!} = \frac{9*10*11*12}{24} = 495.$$

$$32{:}_3 = \frac{32*33*34*35}{(3+1)!} = \frac{32*33*34*35}{24} = 52360.$$

$$32{:}_4 = \frac{32*33*34*35*36}{(4+1)!} = \frac{32*33*34*35*36}{120} = 376992.$$

$$32{:}_5 = \frac{32*33*34*35*36*37}{(5+1)!} = \frac{32*33*34*35*36*37}{720} = 2324784.$$

7.2 Sum of Series of Sequence

Continuous sequence starting from 1 and runs adding 1 in each stage like 1, 2, 3, 4, 5, ... is called as *natural numbers*. Multiple level of Sum of sequence (*Vatsa* with notation $:_v$) of natural number has described on page 93 as:

$$n{:}_v = \frac{n\,(n+1)(n+2)(\ldots n + up\ to\ v+1\,)}{(v+1)!}.$$

What will be the formula in case of subsequent sum of arithmetical sequence (First Series, Second Series, Third Series etc.)

Consider a sequence of 4, 7, 11, ... as below. For our calculation notation, first number of sequence is 'a' and deviation between two consecutive numbers within the sequence is 'd'.

a	3
d	4

EXACT VALUES IN TRIGONOMETRY

n	Sequence	$3^4n:_1$	$3^4n:_2$	$3^4n:_3$	$3^4n:_4$	$3^4n:_5$	$3^4n:_6$
1	3	3	3	3	3	3	3
2	7	10	13	16	19	22	25
3	11	21	34	50	69	91	116
4	15	36	70	120	189	280	396
5	19	55	125	245	434	714	1110
6	23	78	203	448	882	1596	2706
7	27	105	308	756	1638	3234	5940
8	31	136	444	1200	2838	6072	12012
9	35	171	615	1815	4653	10725	22737
10	39	210	825	2640	7293	18018	40755
......				....		....	
31	123	1953	21328	180048	1252152	7466536	39234096
32	127	2080	23408	203456	1455608	8922144	48156240

Recalling $n:_v = \dfrac{n(n+1)(n+2)(....n+\text{up to } v+1)}{(v+1)!}$ of first natural numbers, we use two different Vatsa of any number n.

$$_a{}^d n:_v = n:_{v-1}\, a + (n-1):_v\, d$$

Notation $_a{}^d n:_v$ (spelt as Vatsa sum of sequence up to n numbers having a as the first term having with common deviation d) is sum of sequence as different level.

First Series ($_a{}^d n:_1$) examples: In this case, v is 1, so, v-1 is 0, and $n:_0$ = n; above formula becomes: $n:_{1'} = $ **n a** $+ \dfrac{n(n-1)}{2}$ **d**

$$_4{}^3 5:_1 = \tfrac{5}{1} \times 3 + \tfrac{4*5}{2} \times 4 = 15+40 = 55$$

$$_4{}^3 17:_1 = \tfrac{17}{1} \times 3 + \tfrac{16*17}{2} \times 4 = 51+544 = 595$$

$$_4{}^3 32:_1 = \tfrac{32}{1} \times 3 + \tfrac{31*32}{2} \times 4 = 96+1984 = 2080.$$

This formula is same as common formula for sum of series.

Second Series ($_4{}^3 n:_2$) examples: In this case, v is 2, so, v-1 is 1; above formula becomes: $_4{}^3 n:_2 = n:_1\, a + (n-1):_2\, d$

$$_4{}^3 5:_2 = \tfrac{5*6}{2} \times 3 + \tfrac{4*5*6}{6} \times 4 = 45+80 = 125$$

$$_4 3 17_{:2} = \frac{17*18}{2} \times 3 + \frac{16*17*18}{6} \times 4 = 459 + 3264 = 3723$$

$$_4 3 32_{:2} = \frac{32*33}{2} \times 3 + \frac{31*32*33}{6} \times 4 = 23408.$$

Third Series ($_4 3 n_{:3}$) examples: In this case, v is 3, so, v-1 is 2; above formula becomes: $n_{:3'} = n_{:2}\, a + (n-1)_{:3}\, d$.

$$_4 3 5_{:3} = \frac{5*6*7}{6} \times 3 + \frac{4*5*6*7}{24} \times 4 = 105 + 140 = 245$$

$$_4 3 17_{:3} = \frac{17*18*19}{6} \times 3 + \frac{16*17*18*19}{24} \times 4 = 18411$$

$$_4 3 32_{:3} = \frac{32*33*34}{6} \times 3 + \frac{31*32*33*34}{24} \times 4 = 203456$$

Sum of any number series using above formula are:

$$_4 3 5_{:5} = \frac{5*6*7*8*9}{5!} \times 3 + \frac{4*5*6*7*8*9}{6!} \times 4$$

$$_4 3 17_{:6} = \frac{17*18*19*20*21*22}{6!} \times 3 + \frac{16*17*18*19*20*21*22}{7!} \times 4$$

$$_4 3 32_{:7} = \frac{32*33*34*35*36*37*38}{7!} \times 3 + \frac{31*32*33*34*35*36*37*38}{8!} \times 4.$$

On careful observation of relation between multiplier of "a" and "d" for any number n in $n_{:v-1}\, a + (n-1)_{:v}\, d$:

"a" has multiplied by (i.e. $n_{:v-1}$) given number (n) and its successors to the extent of *Vatsa* (*v*) and divided by factorial of *Vatsa* (*v*).

"d" has multiplied by multiplier to "a" times $(n-1)/(v+1)$

For $_4 3 5_{:1}$: Starting number is 5 and successor multiplier is zero, because we need only one number. The divisor is 1! which is 1. So, multiplier to "a" is 5. This 5 requires to be multiplied by $(n-1)/(v+1)$ means 4/2 which is 10 (=5×4/2) for multiplier of "d". Hence, $_4 3 5_{:1} = 5 \times a + 10 \times d = 5 \times 3 + 10 \times 4 = 55$.

EXACT VALUES IN TRIGONOMETRY

For $_4{}^35_{:2}$: Starting number is 5 and successor multiplier is one number which is 6, because we need only two numbers. The divisor is 2! which is 2. So, multiplier to "a" is 15 (=5×6/2). This 15 requires to be multiplied by (n-1)/(v+1) means 4/3 which is 20 (=15×4/3) for multiplier of "d". Hence, $_4{}^35_{:2}$ = 15×a+20×d= 15×3+20×4= 125.

For $_4{}^35_{:3}$: Starting number is 5 and successor multiplier is two numbers which are 6 and 7, because we need three numbers. The divisor is 3! which is 6. So, multiplier to "a" is 35 (=5×6×7/6). This 35 requires to be multiplied by (n-1)/(v+1) means 4/4 which is 35 (=35×4/4) for multiplier of "d". Hence, $_4{}^35_{:3}$ = 35×a+35×d= 35×3+35×4= 245.

For $_4{}^35_{:5}$: Starting number is 5 and successor multiplier is four numbers which are 6, 7, 8 and 9, because we need five numbers. The divisor is 5! which is 120. So, multiplier to "a" is 126 (=5×6×7×8×9/120). This 126 requires to be multiplied by (n-1)/(v+1) means 4/6 which is 84 (=126×4/6) for multiplier of "d". Hence, $_4{}^35_{:5}$ = 126×a+84×d= 126×3+84×4= 714.

For $_4{}^35_{:7}$: Starting number is 5 and successor multiplier is six numbers which are 6, 7, 8, 9, 10 and 11, because we need seven numbers. The divisor is 7! which is 5040. So, multiplier to "a" is 330 (=5×6×7×8×9×10×11/5040). This 330 requires to be multiplied by (n-1)/(v+1) means 4/8 which is 165 (=330×4/8) for multiplier of "d". Hence, $_4{}^35_{:7'}$ = 330 × a + 165 × d = 330 × 3 + 165 × 4 = 1650.

For $_4{}^35_{:11}$: Starting number is 5 and successor multiplier is ten numbers which are 6 to 15, because we need eleven numbers. The divisor is 11! which is 39916800. So, multiplier to "a" is 1365 (=5 × 6 × 7 × 8 × 9 × 10 × 11 × 12 × 13 × 14 × 15 /39916800). This 1365 requires to be multiplied by (n-1)/(v+1) means 4/12 which is 455 (=1365×4/12). Hence, $_4{}^35_{:7}$ = 1365×a+455×d= 1365 × 3 + 455 × 4 = 5915.

Vatsa of a natural number (n:$_v$) or *Vatsa with sequence* ($_{d}{}^{a}$n:$_v$) of number with deviation (n:$_{v'}$) opens wide range of analytical area in the quantitative analysis in the statistics, financial management, historical analysis of generation, population projection and macroeconomics. Biochemistry analysis, traffic control management (of air space or of networking etc.) or agro-producers' expansion is on *Vatsa series*. It opens new area for nested loops to required infinity level. *Vatsa* of a natural number (n:$_v$) is usable to find the Sine of an angle has described in this book.

Some remarks on Vatsa ($_{d}{}^{a}$n:$_v$):

- Value of n:$_0$ is n.
- In case of first natural number n, the n+6×(n-1):$_2$ is cube of n.
- Sum of cube of n will be equal to squares of n:$_2$.
- In case, $_2{}^1$n:$_2$ is square of number n.
- In case, $_2{}^1$n:$_3$ is sum of square of number n.
- In case, [12× $_2{}^1$n:$_3$ ×n:$_{2'}$] is forth power (n^4) of number n.
- In case, [24× $_4{}^1$(n-1):$_4$ + (n-2):$_4$) + cube of n] is fifth power (n^5) of number n.

***

www.ingramcontent.com/pod-product-compliance
Lightning Source LLC
Chambersburg PA
CBHW071838200526
45169CB00020B/1766